AF902207

Management of Natural Resource for Sustainable Development

The Editors

Dr. B.S. Rathor is currently working as the Head of the Department of Geography, Government Post Graduate College, Morena, M.P. He Graduated in the year 1978 and he had his Post-graduation in 1980 from P.G. College, Ambah, Morena (Jiwaji University, Gwalior). He was awarded with Ph.D. degree in Geography from the Agra University, Agra in 1986. His main area of interest is Physical Geography, Geomorphology of Chambal Ravines and Resource Management. He has jointly published three books.

Dr. Vijay Singh Rathor is working as Assistant Professor of Geography after selected by the M.P. Public Service Commission and ranked top in the merit list. He joined the Government Post-graduate College, Morena.

Management of Natural Resource for Sustainable Development

— *Editors* —
Dr. B.S. Rathor
Dr. Vijay Singh Rathor
Department of Geography,
Government Post Graduate College,
Morena, M.P.

2013
Daya Publishing House®
A Division of
Astral International Pvt. Ltd.
New Delhi – 110 002

© 2013, EDITORS
ISBN 9788170359944

Publisher's note:

Every possible effort has been made to ensure that the information contained in this book is accurate at the time of going to press, and the publisher and author cannot accept responsibility for any errors or omissions, however caused. No responsibility for loss or damage occasioned to any person acting, or refraining from action, as a result of the material in this publication can be accepted by the editor, the publisher or the author. The Publisher is not associated with any product or vendor mentioned in the book. The contents of this work are intended to further general scientific research, understanding and discussion only. Readers should consult with a specialist where appropriate.

Every effort has been made to trace the owners of copyright material used in this book, if any. The author and the publisher will be grateful for any omission brought to their notice for acknowledgement in the future editions of the book.

All Rights reserved under International Copyright Conventions. No part of this publication may be reproduced, stored in a retrieval system, or transmitted in any form or by any means, electronic, mechanical, photocopying, recording or otherwise without the prior written consent of the publisher and the copyright owner.

Published by : **Daya Publishing House®**
A Division of
Astral International Pvt. Ltd
-ISO 9001:2008 Certified Company
4760-61/23, Ansari Road, Darya Ganj
New Delhi-110 002
Ph. 011-43549197, 23278134
E-mail: info@astralint.com
Website: www.astralint.com

Laser Typesetting : **Classic Computer Services,** Delhi - 110 035

Printed at : **Salasar Imaging Systems,** Delhi - 110 035

PRINTED IN INDIA

Prof. N.S. Gajbhiye
PhD., F.N.A.Sc.,
Vice-Chancellor
Doctor Harisingh Gour Vishwavidyalaya
Sagar - 470 003, M.P.

Off: 07582-264796
Res: 223843, 223199
Fax: 07582-264163
E-mail: geakbhiyens@gmail.com
vcsagaruniversity@gmail.com

Foreword

Natural resource management and sustainable development is a prime concern in the wake of large-scale depletion of national resources and extinction of species. In the present sphere of time, when population growth rate is high in the developing countries, over exploitation of the natural resources is also high. This led to depletion and extinction of natural species. Further, imbalances in the utilization pattern paved a way for dividing the societies into haves and have not. The environmental problems ranging from all types of pollution to degradation of land and thus, the climate change are the consequences of population explosion. Food insecurity and malnutrition is therefore creating imbalance in the use pattern of the natural resources.

There is a vital need to use the resources optimally. The development process should be under the scanner of positivity. We have to limit our needs so that the sustainable development could be attended. Framing of policies and implementing them may lead to the sustainable development of any area or region.

This book is an effort to compose all issues that deals with natural resources management. Articles were invited on the various issue of management such as forest resource management, water resources management, soil resources management and all other aspects directly and indirectly related to this matter. It will be useful for the academician, policy planners, research scholars, students and for the public for their wider interest and for the betterment of humanity.

Prof. N.S. Gajbhiye

(Vice-Chancellor)

Preface

Natural resources are the prime bases of the human necessity, power, development and capital. Mankind has been fulfilling his necessary necessities from these natural resources for time immemorial. Resources may be defined as the major sources of human needs. The various forms of human activities such as agriculture, mining, industries, forest produced, fishing and transportation are the consequences of the utilization of the various resources. There is a certain relationship between man and resources. They are interdependent. Although, there are huge reservoirs of the various natural components yet, they cannot be included in the list of resources. Resources are becoming the major sources of the livelihood of mankind due to the proper utilization and this is possible through the human knowledge, efficiency and skill. Therefore the existence of the resources is fully dependent on the men's creativity. Indeed, the physical and mental capability of men, his interest, knowledge, national and social organization, economic development and political stabilities are the human resource that affects the utilization pattern and management of the natural resources.

Natural resources are gifted by the nature which comprises of soil, land, water, natural vegetation, mineral and power resource, wild life and climate - temperature, rainfall and humidity. Earlier, the non renewable resources were abundant because they were underutilized. But now, in the age of science and technology, these resources are over exploited to meet the need of the people and for the development purposes. There is a straight line on the utilization of natural resource in terms of their uses.. Resources are over utilized in the developed countries while in many developing countries; they are underutilized even in the age of globalization, privatization and commercialization. These irregularities in the utilization pattern should be checked immediately to overall development of the nations and its people.

The problems of global temperature and climate changes are a matter of concern for many countries. There were several international meets, on the utilization pattern, sustainable development and climate change took place from time to time and from the Kyoto to Copenhagen. Extreme exploitation and mismanagement of resources given rise to different environmental pollution like land pollution, water pollution, soil pollution, air pollution, sound pollution and mercury pollution.

Sustainable development means long-term development, which is known as continues and constructive development that sustain the need of the present society and also sustain the future generation. It gives rise to the optimum utilization of the natural resources. For conserving and minimum use of the natural resources, reuse, recycle, reduce and regeneration of these resources are vital and the need of the hour.

India is the second largest country in the world from the' point of view to population. Growing population and reducing resources led a way to extinctions of the major resources reservoir. Keeping these issues in mind, we thought to organise a conference and to publish the proceedings of the paper presented in the seminar. This volume is a result of this thinking process and contributions of the conference participants. We take this opportunity to acknowledge the UGC CRO for providing financial assistance. We sincerely acknowledge to the Hon'ble Vice-Chancellor of the Dr. H.S. Gour University, Prof. N.S. Gajbhiye for his contribution. We equally acknowledge Prof. Vijay Singh Tomar, Vice-Chancellor of the RVRS Agricultural University, Gwalior. We are heartily thankful of Prof. R. B. Singh, Delhi University, for his immense contribution from time to time.

Dr. B.S. Rathor

Dr. Vijay Singh Rathor

Contents

Foreword *v*

Preface *vii*

1. **Natural Resin: A Dwindling Forest Resource in the District of Karbi Anglong, Assam: A Study from Geographical Perspective** **1**
Raju Gogoi, Riajur Rahman and B.C. Kalita

2. **Climate Changes** **10**
Priyanka Tiwari and Mona Singh

3. **An Evaluation of Impact of Climate Change on Agriculture with Reference to Indian Agriculture in Particular and World Agriculture in General** **27**
V.V. Barthwal and Pushpa Barthwal

4. **Water Resources Management and Planning for Sustainable Development in Rajasthan** **35**
B.C. Jat and Ram Kumar Gurjar

5. **Groundwater Contamination and its Remediation Strategies: Towards the Challenge of Providing Safe Drinking Water** **54**
Raaz Maheshwari, B.C. Jat and Priyanka Singh

6. **Food and Nutrition Security: The Challenges of Climate Change** **60**
Reeta Mishra, Y.P. Singh, Y.D. Mishra, B.P.S. Raghubanshi and Swati Singh

7. **Effect of Climate Change on Indian Agriculture** **70**
Y.D. Mishra, Reeta Mishra and Swati Singh

8. **Effect of Initial Soil Moisture and Nutrient Management on the Yield of Mustard** **79**
Y.M. Kool, H.S. Bhadoria and A. Singh

9. **The Anthropogenic Costs of Climate Change** **84**
P.B. Reddy and Sateesh Bino

10. **Climate Change Mitigation: What Do We Do** **93**
Rakesh Parmar, Rakesh Sharma and P.B. Reddy

11 **Climate Change and its Impact** **108**
Dilip Gupta

12. **Forest Resources in a Metropolitan City: Delhi** **116**
Chitra Sejwar and B.S. Rathor

13. **Geographical Appraisal of Water Supply of a Metropolitan City: A Case Study of Delhi** **129**
B.S. Rathore and Chitra Sejwar

14. **Climate Change and its Effects on Insects, Pests, Pathogens and Agriculture Environment** **140**
Swati Singh, Y.P. Singh, Dharvendra Singh, K.K. Yadav, Reeta Mishra and S.P.S. Rajput

15. **The Concept of Sustainable Development: Its Needs and Importance** **146**
Satish Modi

16. **Management of Soil Resources for Sustainable Agriculture** **156**
T.R. Rathore and Rajvir Singh Rathore

17. Agricultural Development and Livelihood Sustainability in the Upper Sindh Basin, Madhya Pradesh **164**

Vishwambhar Prasad Sati

18. How Climate Change Affects India **177**

Sunil M. Gawande and Vijay Singh Rathor

19. Natural Resources Utilization and Management for Sustainable Development: A Case Study of Himalayan Watershed **187**

B.L Teli

20. Climate Change Research: Status and Methodology **219**

R.B. Singh and Rajesh Kumar

21. Diversity of Aquatic Animals in National Chambal Sanctuary, Madhya Pradesh **246**

R.K. Sharma, S.C. Bhadoria, B.S. Rathore and N. Dasgupta

22. Climate Change by Global Warming, Ozone Depletion and Acid Rain: The Alarming Disaster for Mankind **262**

Anurag Paliwal

23. Land-Use and Cropping Patterns in a Gwalior District Village Sankhini: A Case Study **268**

M.S. Sisaudiya and Balwant Raj Bhadawar

24. Problem Soils: Their Reclamation and Management **277**

Udai Veer Singh Yadav and P.K. Kulshershtha

25. Cropping Pattern and Crop Combination Analysis for Sustainable Agriculture Development: Buldhana District, Maharashtra **285**

Vijay B. Kharate

26. Forest Management in Morena District **293**

Vinayak Singh Tomar and N.S. Dadoriya

Index 301

Chapter 1

Natural Resin: A Dwindling Forest Resource in the District of Karbi Anglong, Assam: A Study from Geographical Perspective

Raju Gogoi[1], Riajur Rahman[1] and B.C. Kalita[2]*

[1]*Assistant Professor, Department of Geography, Diphu Government College, Diphu, Assam, India*

[2]*Head, Department of Geography, Cotton College, Guwahati, Assam, India*

ABSTRACT

Being one kind of diminishing forest resource, the natural resin used by man for various purposes has been drawing keen attention especially of geographers and botanists for last few decades. It is seen that as soon as Earth's vegetative cover decreases the production of natural resin has also gone down.

The unscientific methods of exploitation of locally available resin from *Canarium resiniferum* and *Canarium bengalense* plants which is famous for its fragrance and other medicinal properties in the hill district of Karbi Anglong, Assam have been found to be responsible for the decrease in production of natural resin.

* Corresponding author: E-mail: raju.dpu@rediffmail.com

Through this research work, it has been tried to find out the real-scene situation of this kind of diminishing forest resource. Analysis has been done from geographical perspectives and efforts have been made to formulate suggestive measures for retrieving the position of availability to meet the future demand of natural resin in the society.

Keywords: *Resin, Karbi, Canarium.*

Introduction

Karbi Anglong, one of the two hill districts of Assam is rich in bio-resources. About 42.70 per cent of the total geographical area of the district is covered with vegetation (in 2009). Large scale of deforestation due to timber smuggling and *Jhum* (Shifting) cultivation has been taking place leading to reduction of many forest resources and disappearance of many valuable tree species. Among them the plants Black Dammar Tree (*Canarium resiniferum/C. strictum*) and East Indian Copal (*Canarium bengalense)* are two of the important species distributed throughout the district which exude a special semi-solid gum-like material from its bark unlike the latex of rubber plants. After getting solidified it becomes natural resin which is commonly known as *'Dhuna'* in Assamese language and *'Hijung'* in local Karbi language. Many other plant species like Sal (*Shorea robusta*), Pine (*Pinus* sps.), Oak (*Quercus* sps.), etc. produce resins of different quantities and qualities in general. The most commonly known resin incense are myrrh, frankincense and copal. But, the resin extracted from *Canarium resiniferum/C. strictum* and *Canarium bengalense* are highly valued for its fragrance and medicinal properties in Assam. It is used as a repellent to mosquitoes, adhesive, at altars during worshipping or in many other traditional and religious occasions, particularly among the Hindus.

The amount of natural resin presently collected through tapping has been found insufficient to meet the market demand in this region. It is observed that a section of the consumers are found today using adulterated resin from the markets.

Objectives

The main objectives of this research paper are: (a) to make an intensive study of the present position of the natural resin in respect of its production and consumption in this part of Assam, (b) to find out and analyze the causes of the present dwindling situation of natural resin production in the hill district of Karbi Anglong of Assam and (c) to suggest possible measures towards the conservation and increase in the number of the source-plant of natural resin to satisfy the need of future generations in this region.

Methodology

This research work is mainly confined to empirical studies. Market observation is followed by extensive field study. Apart from primary data, some related literature and information have been gathered from secondary sources to accomplish the work.

Separate questionnaires have been designed to get information from markets (consumers) and from the resin collectors through random sampling technique. A database has been generated based on the field survey for further studies.

Study Area

Being located relatively in the central part of Assam the hill district of Karbi Anglong, in global context, extends from 25°31′45″ North to 26°36′40″ North latitude and from 92°08′47″ East to 93°53′50″ East longitude. Comprised of two detached physical units of land, the district covers an area of 10,347 square kilometers on the south of the Brahmaputra Valley in the state (Figures 1.1 and 1.2).

Results and Discussion

The field investigation made through the questionnaires among the consumers and collectors of natural resin has given a clear picture on the production and consumption of this resource in the district of Karbi Anglong.

This intensive study reveals that about 70–75 per cent people use natural resin for various purposes like religious faith, repellent to mosquitoes (hygiene), fragrance, freshness, traditional belief and ritual, etc.

Based on quality, the price per kilogram (kg) of the resin in the present retail markets varies from Rs. 100/- to 200/-. Some 10–12 years ago it was just about Rs. 40/-–60/-.

Now-a-days, pure solid resin is hardly available in the market at all the times, whereas adulterated one is easily available.

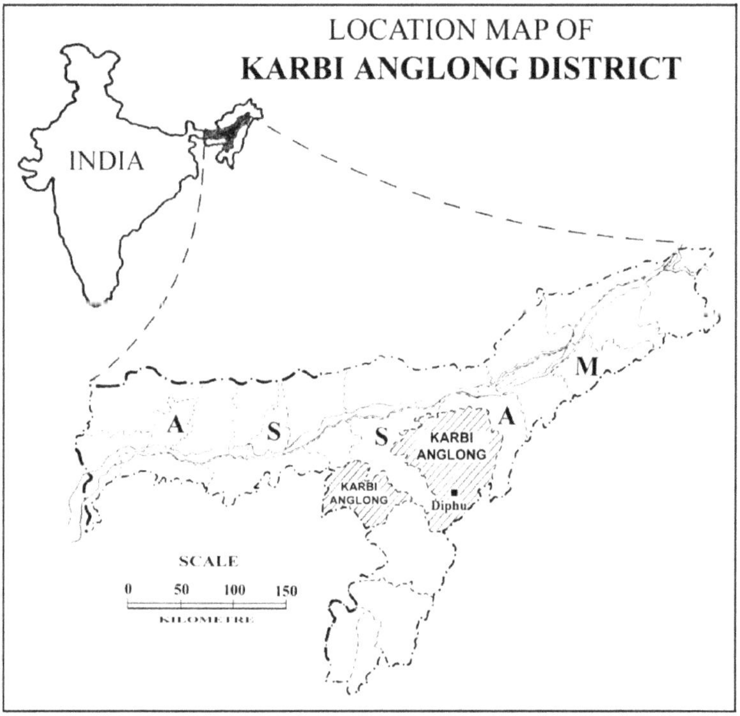

Figure 1.1: Location Map of Karbi Anglong District

Substances like sands, soils, bark of trees, pieces of wood, some gum-like materials collected from other trees having similarity in colours are mixed with the resins. The synthetic resins sold in the markets generally for varnishing purposes are also used some times by the people in place of natural resin which produce bad odour when ignited.

The present study in the district shows about 90 per cent of the consumers surveyed do not have any idea about resin plants. Out of the remaining 10 per cent, 6 per cent people have never seen resin plants nearby. On the other hand, 4 per cent expresses their views that they have seen it in wild state at a far distance.

The survey also encompasses the people in the study area who are adopting resin collection as their partial livelihood. About 98 per cent of the resin collectors collect resins from wild source found

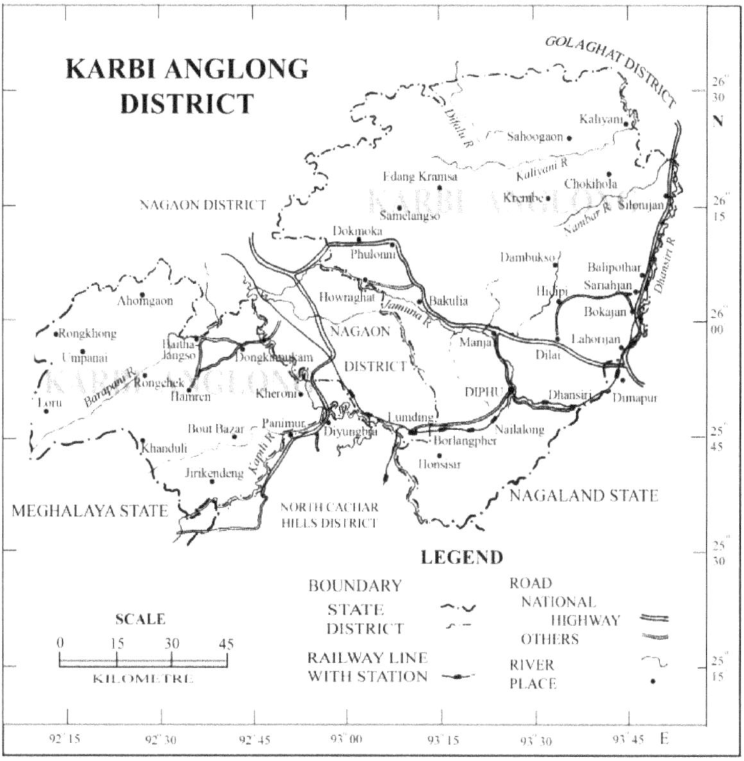

Figure 1.2: Map of Karbi Anglong District

in the Reserved Forests, while only two per cent have resin plants of their own.

Majority of the collectors do not maintain time-based regularity in resin collection. Depending upon the quantities of exudation from plants of various ages they use to collect resins weekly, fortnightly or some times at an interval of one month or more.

Although, there is no fixed amount of resin extracted during their collecting periods, yet, it has been estimated from the study that a collector generally collects an average amount of 10–12 kg per month. The resin collectors during interaction have disclosed that some 20 years back they had a good collection of 40–50 kg resin per month. Some 10–12 years ago the amount decreased to 20–25 kg.

This trend of decline in resin collection is mainly concerned with the decrease in the number of resin plants. Two main reasons have been found responsible for this decreasing trend. First, wholesale destruction of forest land due to traditional practice of *Jhum* cultivation in this tribal dominated hilly district and illegal timber trading. The second and important reason is that most of the collectors have been adopting some unscientific methods of collection. They generally put heavy and deep marks on the trunk of the plants with a *Dao* (machete). Such continuous wounds for long time eventually lead to death of the plants.

Major part of the collected resin is sold in the local markets and a part goes outside the district through some traders. About 30–35 per cent collectors stock resins and sell to regular whole sellers. Other 50–55 per cent collectors sell especially to the local tribal ladies engaged in selling wild fruits, nuts and leafy vegetables in the daily markets. The collectors comprising of 6–7 per cent sell their collected resins by themselves directly in the retail markets. On the other hand, about 3–4 per cent does not have any fixed market; wherever they get opportunity, sell the collected resins.

Based on the way of selling and attitude of the resin collectors, they earn an average amount ranging from Rs. 80/- to 90/- per kg. It is interesting to mention here that most of the tribal ladies hardly follow any measurement of weight; they sell resin by keeping in small parts normally at Rs.10/- each instead.

Therefore, the price of the natural resin in the markets varies from place to place in the district. It is generally high in the urban centres whereas low in the rural markets because of high demand and distant location from the source areas of the earlier.

Empirical observation and interaction with the resin collectors have given a clear picture about the present dwindling situation of the natural resin plants in the study area. At present there is no sufficient number of plants in the forest areas whereas before 10–15 years, it was found in abundance. Due to this situation, most of the collectors have already left this auxiliary occupation and engaged themselves fully in some others.

The Table 1.1 gives a vivid picture about the trend of natural resin production in the district. During the period from 1995–96 to 2001–02 a gradual trend of decline in the production has been observed. It is because of the existence of large association of resin plants in the forest areas in that period. The trend of decline is found to be sharp from the year 2002–03 to 2005–06 due to large scale deforestation. The local ethnic clashes, insurgency problems together with the security checking are also held responsible for the decrease in resin production. The declining trend is again found to be gradual during the period from 2006–07 to 2009–10 though the total production continues to be decreasing. The average amount of decrease within this total period of 15 years has been estimated to be 1,440 kg per annum.

Table 1.1: Production of Natural Resin in Karbi Anglong District, Assam (1995–2010)

Year	*Production (in Kilogram)*	*Year*	*Production (in Kilogram)*
1995–96	39,700	1996–97	39,400
1997–98	39,500	1998–99	38,800
1999–2000	38,600	2000–01	36,900
2001–02	36,500	2002–03	33,100
2003–04	31,700	2004–05	28,000
2005–06	24,200	2006–07	21,300
2007–08	19,500	2008–09	18,700
2009–10	18,100		

Source: Department of Environment and Forest, *Karbi Anglong Autonomous Council, Assam.*

Suggestive Measures

It becomes clear from the above discussion that the present status of natural resin in the district is dwindling. For the conservation and increase in the number of source plants of this forest resource, the following suggestive measures may be adopted.

1. Scientific and proper method of collection of resin without harming the plants is utmost advisable for sustainability and long-term yield.
2. The on going illegal deforestation in the Reserved Forest areas should be prevented.

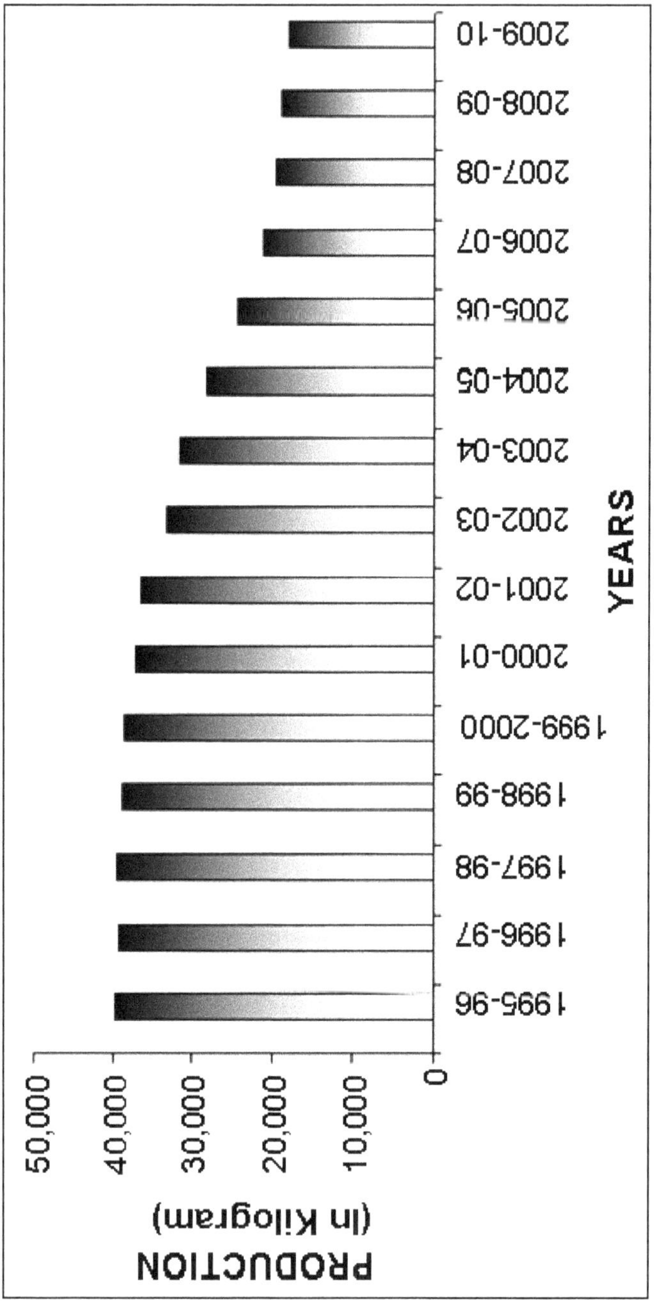

Figure 1.3: Production of Natural Resin in Karbi Anglong District, Assam (1995–2010)

3. A fruitful strategy for resin tree plantation should be adopted through afforestation in the already deforested lands by the concerned authorities. New plantations in private and co-operative sectors should also be encouraged.
4. Selective mother plants of resin should be preserved and the illegal resin collection from the reserved forests should be regularized.
5. Awareness campaigns are to be organized among the people especially the rural tribal in the district.

Conclusion

This valuable forest resource has a great economic significance in this tribal-dominated district. Large scale production of natural resin to meet the present day growing demand will certainly contribute towards economic growth and development of the rural tribal people. In this context, it is to be mentioned here that people's attitude of just using natural resources without thinking for the future generations should be changed, where there lies the concept of *Sustainable Development.*

References

Anderson, K.B., Winans, R.E., Botto, R.E., 1992. The nature and fate of natural resins in the geosphere–II: Identification, classification, and nomenclature of resinites. *Org Geochem.*

Goswami, Dipendra Nath, 2003. The dielectric behavior of natural resin shellac. *Journal of Applied Polymer Science,* John Wiley and Sons, Inc.

Kanjilal, U.N., 1997. *Flora of Assam,* Vol. I, pp. 224–225, Omsons Publications, New Delhi (Reprinted).

Langenheim, Jean, 2003. *Plant Resins: Chemistry, Evolution, Ecology, and Ethnobotany.* Timber Press, Portland.

Mantell, C.L., 1942. *The Natural Hard Resins: Their Botany, Sources and Utilization.* John Wiley and Sons.

Nadkarni, K.M., 1954. *Indian Materia Medica,* Vol. I. Popular Prakashan, Mumbai (Third Revised Edition).

Sharma, Jatindra, 2006–07. *Medicinal and Aromatic Plants of Assam with Special Reference to Karbi Anglong.* Divisional Forest Officer, Silviculture Division, Diphu.

Chapter 2

Climate Changes

Priyanka Tiwari and Mona Singh
Research Scholar (Commerce),
Government T.R.S. College, Rewa, M.P.

Introduction

Business environment can be defined as "the forces, factors and institutions with which the businessman has to deal with to achieve its objectives". In general words we can say business environment is the surroundings in which business exists. Whenever any businessman is operating or working then he has to interact with the customers, suppliers and he has to perform the transactions within the rule and regulations of the government. All these persons, institutions and policies form a part of business environment. Understanding all these factors, persons and institutions carefully is must every businessman. The complete awareness and understanding of business environment is known as environment scanning. Environment scanning is conducted to find out the influence of different factors and persons on the business transactions. Environment scanning can be defined as a process by which organizations monitor their relevant environment to identify opportunities and threats affecting their business.

No company can survive in market by ignoring the effects of business environment. The efficient management analyses the environment and makes changes in organizational policies to integrate its activities with business environment. The most suitable example to prove the impact of business environment is the controversial case of Pepsi and Coke company. Recently a Consumer Organization [(Centre for Science & Environment (CSE)] has claimed to find pesticide contents in the cola drinks of Pepsi and Coke. Although government, health ministry and court have given clean chit to Pepsi and Coke company but then also both the companies are continuously advertising to prove and convince customers that these drinks are safe. They are doing so because the sale and success of these companies not only depend upon internal environment but these depend upon external forces also.

Categories of Business Environment

The business environment can be classified in to two major categories:

- ✰ Economic Environment
- ✰ Non-Economic Environment

1. *Economic Environment*: It consists of factors like fiscal policy, monetary policy industrial policy and economic system of our country.
2. *Non-Economic Environment*: It consists of social, political, legal and technological aspects of business environment.

Features/Characteristics/Nature of Business Environment

The main features of business environment are:

1. *All the external forces*: Business environment includes all the forces, institutions and factors which directly or indirectly affect the business organizations.
2. *Specific and general forces*: Business environment includes specific forces such as investors, customers, competitors and suppliers. Non- human or general forces are Social, Legal, Technological, Political etc which effect the business indirectly.
3. *Inter-relation*: All the forces and factors of Business Environment are inter-related to each other. For example with inclination of youth towards western culture, the demand for fast food is increasing.

4. *Uncertainty*: It is very difficult to predict the change of Business Environment. As environment is changing very fast for example in IT, fashion industry frequent and fast changes are taking place.
5. *Dynamic*: Business environment is highly flexible and keep changing. It is not static or rigid that is why it is essential to monitor and scan the business environment continuously.
6. *Complex*: It is very difficult to understand the impact of business environment on the companies. Although it is easy to scan the environment but it is very difficult to know how these changes will influence business decisions. Some-time change may be minor but it might have large impact. For example, a change in government policy to increase the tax rate by 5 per cent may affect the income of company by large amount.
7. *Relativity*: The impact of business environment may differ from company to company or country to country. For example, when consumers organisation CSE published the report of finding pesticides in cold drinks, resulted in decrease in sale of cold drinks, on the other hand it increased the sale of juice and other drinks.

Importance of Business Environment

The understanding and awareness of business environment is must for every businessman as ignorance about business environment can throw the businessman out from the market. Proper understanding and awareness about the business environment brings lots of benefits for the businessman and these benefits are:

1. *Enables the firms to Identify Opportunities and getting first Mover Advantage*: The businessman who are able to understand and scan the opportunities of business environment at early stage they get maximum benefit or they can capture a big share in the market. They can go much ahead of their competitors. For example, when Maruti Company was planning to sign a contract with Suzuki Company of Japan then the Goodlass Nerolac Company was the first company to understand the environment and to find out that very soon there will be a great demand for

car painting in India and no paint company in our country was having car paint technology. So, Nerolac Company made use of this opportunity and signed a contract with Kansai Paint Company to import the car painting technology and when the Maruti started its car manufacture in India then Nerolac Company got all the car painting contracts of Maruti Company and today also 90 per cent of car paint work of Maruti is done by Nerolac Goodlass Company, by recognizing and understanding the environment at early stage Nerolac Company got the first mover advantage. Afterwards many more paint companies got the technology of car painting but goodlass Nerolac got maximum benefit by recognising the opportunities earliest. They got first mover advantage.

2. *Helps the firm to identify the threats and early warning signal*: The businessmen who are able to scan and understand the business environment on time get a warning signal to deal with the constraints or the negative policies of business environment. Timely scanning of business environment and the qualitative information which businessman get is taken as warning signal and changes in the companies are made. For example, on scanning the business environment, the Maruti Udyog Company got the qualitative information that more foreign car manufacturing companies are going to set up their business in India. They took it is a warning signal and started increasing their production capacity. They increased their production capacity by 3 folds, specially the production of Esteem car which they were manufacturing 65 cars per day increased to 180 cars per day. As a result the sale of Esteem increased to 42,000 cars per year from the earlier sale of 25,400 cars per year. The Maruti company accepted the entry of foreign car companies as warning signal and increased their production to supply the cars without any waiting period to compete with the new companies.

3. *Helpful in tapping and assembling resources*: Businessmen have to supply the goods to market according to the demand in market. To supply output they need input, raw materials etc. They acquire raw materials and other resources

keeping in mind the output demanded in the environment. They select resources according to availability in environment and demand of output in environment. For example, with the demand of flat screen colour T.V. Rather then collecting resources of black and white T.V. They use the scanning report as input to make change in the outlet which suits the taste of present day customer.

4. *Help to adjust and adapt with the rapid changes*: Today changes are taking place very fast and these changes have great impact on business. So, it is essential to understand these change as early as possibly. Business environment scanning helps the companies to scan and understand these changes with environment scanning. Businessmen make change in their internal environment also to match the external environment. So, environment scanning helps in coping with the rapid changes.

   By scanning business environment the Ambani Brothers recognised that today's environment demands quick decision so they shifted from centralisation to decentralisation.

5. *Assisting in Planning and Policy making*: The major strategies or plans and policies in the organisation are formed keeping in mind business environment because the policies and strategies have to be implemented in the presence of environment factors. So, these must be made keeping in mind the environment factors. Scanning of environment factors helps in finding out the opportunities of business and strategies can be made to grab these opportunities. For example, on scanning the Indian Business Environment we find there is great scope for tourism industry in our country. So, businessmen are planning strategies to grab these opportunities. Various hotels of private, public sector, ITC group are making new strategies for tourists. There is a separate travel segment set up and separate leisure and fun segment is set up to get the benefit of upcoming Tourist Industry.

6. *Improvement in Performance*: With continuous scan of Business environment companies can easily improve their performance. By making change in the internal

environment matching to external environment, organisations can prosper and improve their market share. For example, Weston Company which could not co-operate with the changing environment started suffering loss and lost its name in T.V. market whereas BPL, Onida etc. did scan the environment well and are still competing successfully with many multinational companies.

Dimensions of Business Environment

Dimensions of business environment means all the factors, forces and institution which have direct or indirect influence over the business transaction. General Environment is the most important dimension of business environment as businessman cannot influence or change the components of general environment rather he has to change his plans and policies according to the changes taking place in general environment.

Major Components of General Environment

1. Economical Environment

Economic Environment consists of Gross Domestic Product, Income level at national level and per capita level, Profit earning rate, Productivity and Employment rate, Industrial, monetary and fiscal policy of the government etc.

The economic environment factors have immediate and direct impact on the businessman so businessman must scan the economic environment and take timely actions to deal with these environment. Economic environment may put constraints and may offer opportunities to the businessman. After the new economic policy of 1991, lot of opportunities are offered to businessmen. The common factors which have influenced the Indian economic environment are:

(a) Banking sector reform has led to many attractive schemes of deposits and lending money. The Banks are offering loans at very nominal rate of interest and with minimum formalities to be completed.

(b) Recent changes in economic and fiscal policy of country have encouraged NRIs and foreign investors to invest in Indian companies.

(c) Lots of economic reforms are taking place in leasing and financing institutions. The private sector is allowed to enter in financial institutions, as a result customers are gaining.

Some Aspects of Economic Environment

1. Role of Private and Public sector.
2. Rate of growth of GDP, GNP, per Capita Income.
3. Rate of Saving and Investment.
4. Balance of Trade.
5. Balance of Payment.
6. Transport and Communication System.
7. Money Supply in the Economy.
8. International Debt.

2. Social Environment

Social Environment consists of the customers and traditions of the society in which business is exiting. It includes the standard of living, taste, preferences and education level of the people living in the society where business exits. The businessman cannot overlook the components of social environment as these components may not have immediate impact on the business but in the long run the social environment has great impact on the business. For example, when the Pepsi-Cola Company used the slogan of "Come Alive in their advertisement then the people of a particular region mis-interpreted the word "Come Alive" as they assumed it means coming out of graves. So, they condemned the use of the product and there was no demand of Pepsi-Cola in that region. So, the company had to change its advertisement slogan as it cannot survive in market by ignoring the sentiments of the people.

In India also, there are many social reforms taking place and the common factors of Indian Social Environment are:

(a) Demand for reservation in jobs for minority and women.

(b) Demand for equal status of women by paying equal wages for male and female workers,

(c) Demand for automatic machines and luxury items in middle class families,

(d) The social movements to improve the education level of girl child, and

(e) Health and Fitness trend has become popular.

Some Aspects of Social Environment

1. Quality of life
2. Importance or place of women in workforce
3. Birth and Death rates
4. Attitude of customers towards innovation, life style etc.
5. Education and literacy rates
6. Consumption habits
7. Population
8. Tradition, customs and habits of people.

3. Political Environment

Political environment constitutes all the factors to government affairs such as type of government in power, attitude of government towards different groups of societies, policy change implemented by different government etc. The political environment has immediate and great impact on the business transactions so businessman must scan this environment very carefully. The businessman has to make change in his organisation according to the changing factor of political environment. For example, in 1977 when Janata Government came in power they made the policy of sending back all the foreign companies. As a result the Coca Cola company had to close its close its business and leave the country.

The common factors and forces which have influenced the Indian political environment are:

(a) The government in Hyderabad is taking keen interest in boosting I.T. industry, as a result the state is more commonly known as Cyberabad instead of Hyderabad.

(b) After the economic policy of liberalisation and globalisation, the foreign companies got easy entry in India. As a result the Coca Cola which was sent back in 1977 came back to India. Along with Coca Cola, Pepsi Cola and many other foreign companies are establishing their business in India.

Some Aspects of Political Environment

1. Present political system
2. Constitution of the country
3. Profile of political leaders
4. Government intervention in business
5. Foreign policy of government
6. Values and ideology of political parties.

4. Legal Environment

Legal environment constitutes the laws and various legislations passed in the parliament. The businessman cannot overlook the legislations because he has to perform his business transactions within the framework of legal environment. The common legislation passed which has affected the business transactions are Trade Mark Act, Essential Commodity Act, Weights and Measures Act etc. Most of the time legal environment put constraints on the businessman but sometimes they provide opportunities also. The common instances of Indian legal environment which have influenced business transactions recently are:

(a) Deregulation of capital market have made it easy for businessman to collect capital from primary market.
(b) Removal of control from the foreign exchange and liberalisation in investment is encouraging foreign investors and NRIs to invest in Indian capital market.
(c) Advertisement of alcoholic product is Prohibit.
(d) Compulsory to give statutory warning in Tobacco production.
(e) Delicencing policy of industries.

Some Aspects of Legal Environment

1. Various laws and legislative acts
2. Legal policies related to licensing
3. Legal policies related to foreign trade
4. Statutory warnings essential to be printed on label
5. Foreign Exchange Regulation and Managements.

5. Technological Environment

Technological environment refers to changes taking place in the method of production, use of new equipments and machineries to improve the quality of product. The businessman must closely monitor the technological changes taking place in his industry because he will have to implement these changes to remain in the competitive market. Technological change always bring quality improvement and more benefits for customers. The recent technological changes of Indian market are:

(a) Digital watches have killed the prospects and the business of traditional watches.

(b) Colour T.V. technology has closed the business of black and white T.V.

(c) Artificial fabric has taken the market of traditional cotton and silk fabrics.

(d) Photo copier and xerox machine have led to the closure of carbon paper business.

(e) Shift in Demand from vacuum tubes to transistors.

(f) Shift from steam locomotives its diesel and electric engine.

(g) From typewriter to World Processors.

Some Aspects of Technological Environment:

1. Various Innovations and Inventions
2. Scientific Improvements
3. Developments in IT sector
4. Impact and Export of Technological
5. Technological Advances in Computers.

Economic Environment in India

Economic Environment is the most important dimension of Business Environment. The important factors of economic factors which have impact on Businessmen are:

1. *Stage of Economic Development*: The Business policies of a developed country are different from the policies of developing countries.

2. *Economic Structure*: There are mainly three economic structures Capitalist, Socialist and Mixed Economic structure. In capitalist state businessman takes decisions by giving priority to profit motive. In socialist state government give in priority to welfare of people and in mixed economy *i.e.*, India Business Policies are taken by considering both *i.e.*, profit and welfare of people. In mixed economy there is co-existence of Public and Private sectors.
3. *Economic Policies*: Important economic policies which influence Business decisions are Industrial Policy, Fiscal Policy, Monetary Policy. Liberal Policies offer more opportunities to businessman whereas strict policies put constraints.
4. *Economic Planning*: It includes five years plan, Annual Budgets, Growth plan etc. These plans also influence business decisions.
5. *Economic Indicators*: The common economic indicators are National Income, Per Capital Income, Rate of Saving and Investment, Balance of Payment, Value of Export Import, GDP, GNP etc.
6. *Infrastructure*: It refers to basic services or setup necessary to carry on activities in a country *i.e.*, Transportation, Communication, Banking, Financial Institution etc.

Economic Environment in India since Independence

There has been tremendous change in Indian economic environment since Independence. The main features of Indian Economic Environment at the time of Independence were:

1. Dominance of Agriculture Sector.
2. Almost three-fourth of total population was occupied in Agriculture Sector.
3. 85 per cent of the population was living in villages.
4. Use of obsolete and outdated technology in production.
5. Lack of good public health system.
6. Widespread communicable diseases.

Immediately after Independence in order to solve the economic problems, government of India made some plans and policies. The main features of these plans and policies was giving importance to Public sector. Many areas were reserved only for public sector and curtaining the role of Private sector, Many restrictions were put on Private sector.

The main objectives of new Development Plans were:

1. Fiscal Deficit reached 7 per cent (approx.) of GDP.
2. Internal Dept rose to 50 per cent of GDP.
3. Negative Agricultural growth.
4. Inflation Rate rose to 13-14 per cent.
5. Depreciation in the value of Rupee.
6. Fall in foreign exchange reserve.
7. Fall in foreign exchange reserve.
8. Credit-worthiness of International financial institutions of our country fell.
9. Negative Balance of Payment.
10. India was on the verge of becoming defaulter. As a result SBI and sold and pledged gold in international market.

To save country from the serious situation of Government of India took some reform measures.

Reform Measures

1. Reducing fiscal deficit.
2. Announcement of New Industrial Policy in July 1991.
3. Abolition of Industrial Licensing Policy.
4. Amendment of MRTP Act.
5. Open entry for Private sector to areas which were earlier reserved for Public sector only.
6. Rise in foreign equity holders from 40 per cent to 51 per cent.
7. Setting up of Foreign Investment Promotion Board (FIAB)
8. Introduction of India Development Bond scheme to get funds from abroad.
9. To check the devaluation of rupee a stand by credit was negotiated from IMF.

10. Buying back of gold which was earlier pledged.
11. Measures to control import and encourage export.
12. Introduction of Liberalized Exchange Rate Management System (under this there was dual exchange rate)
13. Elimination of import license on capital goods.
14. Abolition of Export duty.

The major and most important measure taken by government as economic reform was announcement of New Industrial Policy in July 1991.

The Features of New Economic Policy 1991

The main characteristics of new Economic Policy 1991 are:

1. *Delicencing*: Only six industries were kept under Licensing scheme.
2. *Entry to Private Sector*: The role of public sector was limited only to four industries; rest all the industries were opened for private sector also.
3. *Disinvestment*: Disinvestment was carried out in many public sector enterprises.
4. *Liberalisation of Foreign Policy*: The limit of foreign equity was raised to 100 per cent in many activities *i.e.*, NRI and foreign investors were permitted to invest in Indian companies.
5. *Liberalisation in Technical Area*: Automatic permission was given to Indian companies for signing technology agreements with foreign companies.
6. *Setting up of foreign Investment Promotion Board (FIPB)*: This board was set up to promote and bring foreign investment in India.
7. *Setting up of Small Scale Industries*: Various benefit were offerd to small scale industries.

The major changes or impacts of the new economic policy 1991 are:

- ✰ Liberalisation
- ✰ Privatisation
- ✰ Globalisation

1. Liberalisation

Liberalisation refers to end of licence, quota and many more restrictions and controls which were put on industries before 1991. Indian companies got liberalisation in the following way:

(a) Abolition of licence except in few.

(b) No restriction on expansion or contraction of business activities.

(c) Freedom in fixing prices.

(d) Liberalisation in import and export.

(e) Easy and simplifying the procedure to attract foreign capital in India.

(f) Freedom in movement of goods and services.

(g) Freedom in fixing the prices of goods and services.

2. Privatisation

Privatisation refer to giving greater role to private sector and reducing the role of public sector policy of privatisation government took the following steps:

(a) Disinvestment of public sector *i.e.,* transfer of public sector enterprise to private sector.

(b) Setting up of Board of Industrial and Financial Reconstruction (BIFR). This board was set up to revive sick units in public sector enterprises suffering loss.

(c) Dilution of Stake of the Government. If in the process of disinvestment private sector acquires more than 51 per cent shares then it results in transfer of ownership and management to the private sector.

3. Globalisation

It refers to integration of various economies of world. Till 1991 Indian government was following strict policy in regard to import and foreign investment in regard to licensing of imports, tariff, restrictions etc. but after new policy government adopted policy of globalisation by taking following measures:

1. *Import Liberalisation*: Government removed many restrictions from import of capital goods.

2. Foreign Exchange Regulation Act (FERA) was replaced by foreign exchange management Act (FEMA).
3. Rationalisation of tariff structure.
4. Abolition of Export duty.
5. Reduction of Import duty.

As a result of globalisation physical boundaries and political boundaries remained no barriers for business enterprise. Whole world become a global village.

Globalisation involves greater interaction and interdependence among the various nations of global economy.

Impact of Change in Economic Policy on the Business or Effects of Liberalisation and Globalisation

The factors and forces of business environment have lot of influence over the business. The common influence and impact of such changes in business and industry are explaining below:

1. *Increasing Competition*: After the new policy, Indian companies had to face all round competition which means competition from the internal market and the competition from MNCs. The companies which could adopt latest technology and which were having large number of resources could only survive and face the competition. Many companies could not face competition and had to leave the market. For example, Weston Company which was a leader in T.V. market with more than 38 per cent share T.V. market lot its control over the market because of all round competition from MNCs. 1995-96, the company almost became unknown in the T.V. market.
2. *More Demanding Customers*: Prior to new economic policy there were very industries or production units. As a result there was shortage of product in every section. Because of this shortage the market was producer-oriented *i.e.*, producers became key in the market. But after new economic policy many more businessman joined the product line and various foreign companies also established their production units in India. As a there was

surplus of products in every sector. This shift from shortage to surplus another shift in the market *i.e.,* producer market to buyer market. The market customer-oriented and many new schemes were made by companies to attract the customers. Now a days products are produced/ manufactured keeping in mind the demands of the customers.

3. *Rapidly Changing Technological Environment*: Before or prior to new economic there was a small internal competition only. But after the new economic policy the world competition started and to stand this global competition the companies need to adopt world class technology. To adopt and implement the world class technology the investment R&D department has to increase. Many pharmaceutical companies increased their investment in R and D department from 2 per cent to 12 per cent and companies started spending a large amount training the employees.
4. *Necessity for Change*: Prior to 1991 business enterprises could follow stable a long period of time but after 1991 the business enterprises have to modify their policies operation from time to time.
5. *Need for Developing Human Resources*: Before 1991 Indian enterprises were by inadequately trained personnel's. New market conditions require people with competence skill and training. Hence Indian companies felt the need to develop their skills.
6. *Market Orientation*: Earlier firms were following selling concept *i.e.,* produce first then go to market but now companies follow marketing concept *i.e.,* planning production the basis of market research, need and want of customer.
7. *Loss of Budgetary Support to Public Sector*: Prior to 1991 all the losses of Public sector were used to be nade good by government by sanctioning special funds from budgets. But today the public sectors have to survive and grow by utilizing their resources efficiently other-wise these enterprises have to face disinvestment. On the whole the policies of liberalization, globalization and privatization

have brought positive impacts on Indian business and industry. They have become more customer focus and have started giving importance to customer satisfaction.

8. *Export a Matter of Survival*: The Indian businessman was global competition and the new trade policy made the external trade very liberal. As a result to earn more foreign exchange many Indian companies joined the export business and got lot of success in that. Many companies increased their turnover more than double by starting export division. For example, the Reliance company, Videocon, MRF, Ceat Tyres etc. got a great hold in the export market.

Chapter 3

An Evaluation of Impact of Climate Change on Agriculture with Reference to Indian Agriculture in Particular and World Agriculture in General

V.V. Barthwal[1] *and Pushpa Barthwal*[2]

Lecturer in E.A.F.M., Government Lohia PG College, Churu

Lecturer in Hindi, Government College, Sardarshahar

ABSTRACT

In this chapter an attempt has been made to analyze the impact of climate change on Indian agriculture in particular and world in general. Climate change refers to significant variation in either the mean state of the climate or in its variability, persisting for an extended period, typically, decays or longer. It is evident that there was, there is, and there will be climate variability of global, regional and local levels. Since, climate is closely related to human activities and economic development including agricultural system, there is serious concern about it's stability. Uncontrolled human activities like burning fuel, deforestation, various agricultural system and industrial practices are altering the composition of atmosphere and contributing to it. These activities produce greenhouse gases, which are harmful to atmosphere. It assists in increase of temperature in the environment as well as in earth's temperature. It is estimated that just a few degrees of increase in the earth's temperature can cause draughts and crop failures, ecosystem imbalances and melting ice caps causing sea levels to rise. This phenomenon of climate change is hampering the world's agriculture in general and India in particular. Study points our that between 1996 and 2003; grain production has stabilized slightly over 18000 tons. In, 2000, 2001, 2002 and 2003 grain stock have been dropping, resulting in a global grain harvest that was short of consumption by 93 millions of tons in 2003. In, 2002, India and United State suffered sharp harvest reductions because of record temperature and droughts. Based on the impact of greenhouse gases already in the atmosphere, changes in climate are expected to lead a drastically reduced agriculture production in the world, with potentially a 50 per cent reduction in Africa alone. The reduction will affect global species and disproportionately impact on the poor. In this connection National Mission o Strategic Knowledge for climate change (India) has rightly pointed out that the impact of climate change could prove particularly severe for women. With climate change, there would be increasing scarcity of water, reduction in yields of forests biomass and increased risks to human health with children, women and vulnerable. With the possibility of decline in food grains, the threat of malnutrition may also increase. All these would add to deprivations that a women already encounters and so special attention should be paid to the aspect of the gender. Again it also highlights that food production in India is sensitive to climate changes such as variability in monsoon, rainfall and temperature changes within a season. Studies by Indian Agricultural Research Institute (IARI) and others indicate greater expected reduces the *Rabi* Crop. Every 1°C rise in temperature reduces wheat production by 4-5 Million Tones. Small changes in temperature and rainfall have significant effects on the quality of fruits, vegetable, tea, coffee, aromatic and medical plants and basmati rice. Pathogens and insect populations are strongly depended upon temperature and humanity, and changes in these parameters may change their population dynamics. Other impacts on agricultural and related sectors include lower yields from dairy, cattle and decline in fish breeding, migration and harvests. Global reports indicate a loss of 40 per cent in crop-production by 2100.

Introduction

It is well known fact that uncontrolled human activities and resource consumption is affecting adversely to the environment of this planet. Today human community has been facing so many ecological and environmental problems which are big threats for the survival of the mankind, some of them are climate change, environmental degradation, sustainable development, global warming and food security for growing population climate change is primarily caused by the building up of greenhouse gases (GHG) *e.g.*, Carbon dioxide, methane, nitrous oxide and other in atmosphere. The global increase in carbon dioxide concentration which have been responsible for the largest part of this buildup, are due primarily to fossil fuel use and land use change, owing to human activities taking place since pre-industrial times, while those of methane and nitrous oxide are primarily due to agriculture. Thus climate change or global warming is caused by the release of greenhouse gases in to the atmosphere. These gases accumulate in the atmosphere and increases the effect of relative forcing on the climate change resulting in a warming of the atmosphere. The change in greenhouse gas concentration are projected to lead to regional and global changes in climate and weather parameters such as temperature, precipitation soil, moisture and sea level. It is observed that during last 100 years rise in temperature is up to about 0.72°C. The Intergovernmental Panel on Climate Change (IPCC) has rightly mentioned in it's fourth assessment report that "the warming of the climate system is now unequivocal, as is now evident from observations of increase in global average air and ocean temperature,wide spread melting of snow and ice and rising global sea levels. According to panel's report (1996), as increase in atmospheric concentration of greenhouse gases equivalent to a doubling of carbon dioxide (CO_2) will force a rise in global average surface temperature of 1.0 to 3.5 degree Celsius by 2100.

Agriculture Climate Change

Every increase in temperature affects the cropping pattern and sowing period. Various studies indicate that agriculture is sensitive to climate variability change which contribute to climate change and also will be affected by climate change. Agriculture productivity can be affected by climate change in following ways:

1. Directly due to change in temperature, precipitation and/ or CO_2 levels.
2. Secondly- indirectly, through change in soil distribution and frequency infestation by pests, insects diseases or weeds.
3. A cute water storage condition combined with thermal stress, could adversely affect wheat and more severely rice productivity in India ever under the positive effect of elevated CO_2 in future.

The agriculture will be severely suffered by climate change with the impacts of climate change felt more frequently in various parts of the world in the form of severe droughts, water, shortages and extreme weather events, the time for speculation over the causes and consequences of climate change is over. The IPCC found that there is abundant evident that freshwater resources on which the viability of agriculture depends, are vulnerable and have the potential to be strongly impacted by climate change, and that current water management practices may not be sufficient to cope with these impacts. (IPCC-2001) More frequent extreme weather conditions, stresses on water availability, droughts and overall changing environment conditions all pose serious threat to agriculture production. The FAO projects a reduction in global cereal output in 2009 due to smaller plantings and/or adverse weather in most of the world major producers (FAO, 2009). Increasing temperature will impact agricultures production, higher temperature reduces the total duration of a crop cycle by inducting early flowering, thus shortening the "grain till" period. The shorter the crops cycle the lower the yield per unit area. Increased temperature also mean increased evaporation and transpiration rates even a small increase of 1°C could increase the rate of evaporation by 5 to 15 per cent. In north India, for instance, a temperature rise of 0.5°C could reduce wheat yields due to heat stress by about 10 per cent, if rainfall does not increase. It is predicted by several scientists that a temperature increase of 3°C will result in a 15 to 20 per cent decrease in wheat yields and also a decrease in rice yields rise in surface temperature will create more conductive conditions for pest infection, which is already a major constraint in achieving higher crop production in India, and hence loss of crop.

The following are some of the negative influence of climate change, could adversely affect agricultural production resulting to:-

1. The yields of different crops and geographic limits may be altered by change in soil moisture, temperature, precipitation, cloud cover as well as in CO_2 concentration.
2. The change in soil properties such as the as the loss of soil organic matter, leaching of soil nutrients, salinizations and erosion are a likely consequences of climate change for some soils in some climate change.
3. The effect of climate on pests may add to the effect of other factors such as the overuse of pesticides and the loss of biodiversity, which already contribute to plant pest and disease outbreaks.
4. Flooding will probably become a significant problem in some already flood prone regions of Asia such as China and further south in Eastern Asia.
5. Decrease in productivity are most likely in these regions, which are already flood-in secure.
6. The summer monsoon is predicted to become stronger and move north-west ward. However, this increased rain could be beneficial to some areas in addition to change in the frequency of extreme climatic events, changes in rainfall and temperature could be damaging and costly to agriculture.
7. The range of many insects will change or expand and new combinations of diseases and pests may enough as natural ecosystems responds to shifts in temperature and precipitation profiles.
8. The climatic change due to global warming is also affecting agricultural productivity. Table 3.1 exhibits minimum and maximum yield variability achieved in food grains in the country due to various factors.
9. Intensive agriculture due to higher application of fertilizer and pesticides and over irrigation especially in Indo-Gangetic plains is causing severe environmental degradation. This is affecting the agricultural productivity.

Table 3.1: Diversities in Yield of Food Grains (2007-08)

Crops	*Minimum State Yield (Kg/ha)*	*Maximum State Yield (Kg/ha)*
Food Grains	Rajasthan, 1119	Punjab, 4017
Wheat	Karnataka, 762	Haryana, 4232
Rice	MP, 824	Punjab, 3868
Coarse Cereal	Chhattisgarh, 537	AP, 3033
Pulses	Orissa, 445	Haryana, 824
Oilseeds	Assam, 494	TN, 1829

Source: GOI–(2009-10)Economic Survey–Ministry of Economic Affairs New Delhi.

Conclusion and Suggestions

In summing up, food production India is sensitive to climate changes such as variability in monsoon rainfall and temperature changes with in a season. In this connection, studies by Indian Agricultural Research Institute (IARI) and others indicate greater expected loss in the *Rabi* crop. Every 1°C rise in temperature reduces wheat production by 4-5 Million tones. Small changes in temperature and rainfall have significant effectts on the quality of fruits, vegetable, tea, coffee, aromatic and medical plants and basmati rice, other impacts on agricultural and related sectors include lower yield from dairy cattle and decline in fish breeding, migration and harvest. Global reports indicate a loss of 10-40 per cent in crop production by 2100. It is revealed that climate change and and agriculture are interrelated process, both of which take place on global change. People living in poverty, small and marginal farmers practicing agriculture on rainfed farms will bear the brunt of climate change. According to Indian Council of Agricultural Research, medium term climate change predictions have projected the likely reductions in crop yields due to climate change at between 4.5 per cent by 2039. The long run predictions points a scarier picture with the crop yields anticipators to fall by 25 per cent or more by 2099. Thus will have a deterimental affect on farmers income and purchasing power, with obvious down–the–line repercussions. Hence the mitigation of climate change is urgent need of hour and it should be addressed very wisely. In this connection use of technological advances can play their vital role to combat climate change, there has to be sound policy frame work and strong political will to achieves this objectives. State agriculture universities and regional farm research centers, will have to play a role in developing local situation specific for adopting the rain fed farming to emerging climate patterns. In this connection following steps can be taken to mitigation of climatic change:

1. Environmental laws should be implemented properly and effectively, *i.e.* Environment (Protection) Act. 1986.
2. The farmers may have to be trained to adopt conservation technology to reduce negative trend of organic carbon depletion by increasing the use of organic and bio-fertilizer and use balanced dose of NPK (4:2:1) instead of present practice as observed in few states as high as 75:24:1.
3. Integrated pest management is eco-friendly and thus a balance between pest control technology need to be pursued. Use of bio-pesticides should be increased to reduce the pollution of water by pesticides. Industries effluent needs to be properly treated before discharging in to rivers through strict regulation.
4. There is a tremendous scope to use renewable energy to supplement the agricultural energy demand. Bio-fuels have added advantage of lower GHG emission. Bio-diesel produced from plant origin material's can be well as in tractors. Biogas, producer gas, solar photo vatican energy wind energy are other sources which have successfully been used in agriculture and industrial application.

References

Kumar, K.S. Kavi, 2007. Climate change studies in Indian agriculture. Economic and Political Weekly, November–17, pp. 13–18.

Kumar, K.S. Kavi, 2009. Climate sensitivity of Indian agriculture. In: *South Asian Network for Development and Environmental Economics (SANDEE),* Working Paper No. 45–09, Madras School of Economics, Chennai November.

Mendelsohn, R.W. Nordhaus and Show, D., 1994. The impact of global warming on agriculture a Ricardian analysis. *American Economic Review,* 84(4): 753–771.

Ninawe, A.S., 2010. Climate change: Global impact on agriculture. Fisheries and Animal Husbandry, *University News*, 48(24), June 14–20.

Rashmi, R.R. and Satapathy, S., 2010. Climate change facing the challenge. *Yojana*, April.

Reddy, R. Raghava, 2010. Impact of weather extreme on Indian agriculture in the context of climate change. *University News*, 48 (24), June 14–20.

Singh, Gyanendra, 2010. Indian agriculture and climatic change: Issues and challenges. *University News*, 48(24) June 14–20.

Soloman, S.D., Quin, M. Manning *et al.*, 2007. *Climate Change 2000: The Physical Science Basis*. Contribution of Working Group I to the Fourth Assessment Report of IPCC, Cambridge, Cambridge University Press.

Chapter 4

Water Resources Management and Planning for Sustainable Development in Rajasthan

B.C. Jat[1] *and Ram Kumar Gurjar*[2]

[1]*Department of Geography, Government P.G. College, Neemkathana, Sikar, Rajasthan*

[2]*Assistant Professor, Department of Geography, University of Rajasthan, Jaipur, Rajasthan*

Introduction

Water has been an important symbol in the legends and histories of many ancient cultures. Unlike people living in the urban, industrialized nation of 21st century, most humans through out history knew that their water resources could run out, and they developed a healthy respect for conserving whatever water they found. Suddenly it is so clear: India is running out of freshwater. Humanity is polluting, diverting, and depleting the well spring of life at a starting rate with every passing day. Our demand for freshwater out poces its availability and thousands more people are put at risk. Already, the social, political and economic impacts of water scarcity are rapidity becoming a destabilizing force, with water related conflicts springing up around the country. Quite simply, unless we dramatically change our ways, between one-half and two-third of humanity will be living with severe freshwater shortages within the next quarter century.

The trends of water scarcity and water use that we witness currently in the Rajasthan have prompted the planners, the administrators, researchers and intellectuals to ponder over the problem which may become very acute for the people of the twenty-first century and that day is not very far off. This problem will not be confined to Rajasthan alone and if it is not dealt with satisfactory, it will make life difficult on this planet and may end into a tragedy for the human race. This evil trend is man's own creation because he had indulged into indiscriminate exploitation of water causing great imbalance of the ecosystem.

The problem of water crisis can not be solved without people's participation and efforts should be made to get every citizen involved at the different water conservation schemes. The present work is a effort in this direction and suggests various indigenous water harvesting techniques which will be useful for better water management system. Ecologist, political elite, policy planners, irrigation technocrats, agronomic scientists, social scientists interested in ecological parameters of developmental change and all others interested in water management would find this research a useful reservoir of descriptive and technical touches focused on a fragile eco-system undergoing far-reaching changes undertaken without much fore-thought.

Study Area

The state Rajasthan is situated in north western part of India and extends from 23°3' North to 30°12' North latitudes and 69°30' East to 78°17' East longitudes. It covers an area of 3,42,239 sq.km. Physiographical the state has four following distinct units:

1. Western sandy plains and sand dunes,
2. Aravalli hill ranges and adjoining areas,
3. Eastern plains, and
4. Vindhyan scarp land and Deccan Lava plateau. The physical features of study area are dominated by the low rugged highly dissected topography. There are different hydrogelogical environments in which ground regime behaves differently, which has to be considered while

planning its exploitation of groundwater a port from this there are problems of considerable depletion of water table groundwater quality and pollution. Surface water flow has been much influenced by its relief and climate.

Rajasthan having mainly arid to semi arid climate conditions and meager surface water resources. The climate of Rajasthan is influenced deeply by the division of the state into two distinct parts by the Aravalli ranges. Although the whole state experiences hyperthermic conditions, the areas west of the Aravallis are very hostile, with great extremes of temperatures, long period of severe droughts, high wind velocity and low humidity. The areas to the east of the Aravallis experience less stressing climate, although there are great variations in rainfall and temperature. Rainfall is the main source of all the surface and subsurface water. The rainfall trend over the state is highly erratic in time and space causing floods and draughts simultaneously at one or other places due to its geographical setting and distance from Ocean. the monsoon rainfall over the State depends on south-west monsoon which extends from second week of June till September. Most of the rainfall is experienced during monsoon season.

The annual rainfall in the state varies significantly from 164 mm in Jaisalmer to 1004 mm in Jhalawar, with a mean annual rainfall of 573 mm. The general trend of isohytes is from the northwest to the south-east.

The average annual rainfall in this part ranges from less than 100 mm in northwestern Jaisalmer (lowest in the state) to 200 to 300 mm in the Ganganagar, Bikaner and Barmer regions, 300 to 400 mm in the Nagaur, Jodhpur, Churu and Jalore regions and more than 400 mm in the Sikar, Jhunjhunu and Pali regions, and along the western fringes of the Aravallis. On the eastern side of the Aravallis, the rainfall varies from 550 mm in Ajmer to 1020 mm in Jhalawar. In the plains, Banswara (920 mm) and Jhalawar (950mm) districts receive the maximum annual rain.

The yearly total rainfall is hightly variable all over the state. It is most erratic in the western half, with frequest spells of drought, punctuated occasionally with heavy downpours in some years, due to low-pressure systems travelling over the region.

During the last decade *i.e.*, (1999-2000) it had been observed that from 1990 to 1994, alternate wet and dry phases have been experienced over the State followed by wet spells up to 1997, whereas in 1998, it was significantly below normal. In the year 1999 and 2000 the State received rainfall far below normal. As a result the State had faced draught condition.

The average relative humidity in Rajasthan is 60 per cent -65 per cent, being at a low of 50 per cent in the west to above 70 per cent in eastern Rajasthan.

Methodology

The study is based on primarily secondary data collected from various government and non government offices and organizations. State groundwater board, Jaipur, central groundwater board, western office, Jaipur, state irrigation department Jaipur, state agriculture department, Jaipur, state department of PHED, planning, forest, horticulture are major offices from where secondary data have been gathered.

Village level and household samples at micro level have been generated to primary data for evaluation of water management. the present study is based on the available secondary data base for the preparation of an initial base profile for present and future sources of water.

Water Resources Position in Rajasthan

Rajasthan with about ten per cent of country's geographical area and 5.5 Per cent Population is endowed with about one per cent of its total water resource which evidently speaks of the position of this crucial commodity in the state all management and planning of water in the state should keep this pivotal fact in focus and provide priority significance on merit basis to various user sectors of its economy, taking spatial variations into account while charting developmental policies plans and strategies.

Rainfall, the major source of water supply is highly deficient, variable and unpredictable in its reliability throughout the state. Impressionistic views reveal and emphasize decline in the overall rainfall, increase in its variability in different areas, marked fluctuations in seasonal behaviour pattern. A closer, comprehensive and critical analysis of the availably rainfall records and data of district, tehsil and other stations can enlighten further on this spatio-temporal phenomena and base for futuristic planning.

Groundwater Resource

Groundwater potential area in the state is 2,15,142 sq.km. which is 63 per cent of State, whereas, saline area covers 100223 sq. km. mainly falling in western Rajasthan 29 per cent of State. Hilly terrain covers and area of 26874 sq. km. (8 per cent of State).

Groundwater resources in the State have been computed as up to 1.1.98 on the basis of Groundwater Estimation committee guidelines. Accordingly gross annual recharge, groundwater draft, balance and stage of groundwater development have been computed. Groundwater recharge, draft and balance, as computed is 12602, 8708 and 3894 mom respectively. Overall stage of groundwater development comes to 69.10 per cent. Categorisation of blocks has been done on the basis of stage of stage of state of groundwater development and long term trend of water levels during pre-monsoon and post-monsoon period.

Table 4.1: Criteria of Categorisation

Category	*Groundwater Withdrawal*	*Trend of Water Level*
Safe (white Zone)	less then 70 per cent	No significant depletion water level
Semi critical (Grey Zone)	70-90 per cent	Water level depletion either in pre- or post-monsoon period.
Critical (Dark Zone)	More then 90 per cent	Water level depletion either in pre- or post-monsoon period.
Over exploited	More then 100 per cent	Water level depletion in both pre- and post-monsoon period.

Based on above criteria, 237 blocks of State has been categorized as under:

Sl.No.	*Category*	*No. of Blocks*	
		2000	*2005*
1.	Safe	135	50
2.	Semi critical	34	21
3.	Critical	26	80
4.	Over exploited	42	86
		237	237

Groundwater level is controlled by its location. Generally, in flat and alluvial areas, water levels are deep, whereas in the valley fills, it is shallow. In hard-rock areas, the effective thickness of the fractured zone usually does not exceed 25 to 30 metres below land surface; therefore, water levels are fairly shallow. Similarly, in the vicinity of surface water bodies the depth of water is shallow. sometimes, water logging problems develop due to the rise of groundwater levels near or above the land surface in canal command areas. areas with thick alluvial deposits and low rainfall usually exhibit deep water level conditions.

The depth of water level below land surface varies widely throughout Rajasthan. Shallow water levels of less than 10-20 metres are recorded in the command area of the Indira Gandhi Nahar Pariyojna in parts of Ganganagar, Hanumangarh and Bikaner districts, the Mahi command in Banswara and Dungarpur districts, and the Chamabal command in Kota, Bundi and Baran districts. East of the Aravallis, the water levels are generally less than 10 metres to 25 metres below land surface. Moderately deep water levels of 20-40 meters below land surface are seen in Pali, Jalore and major part of Barmer, Jodhpur, Nagaur, Sikar and Jhunjhunu districts. Further west and northwest in Jaisalmer, Bikaner, and parts of barmer, Jodhpur, Nagaur, Churu and Sikar, water levels are usually 40-80 metres deep. A few pockets in Jaisalmer and Bikaner have groundwater level more than 100 metres deep.

Groundwater Level

Water levels very with rainfall, physiography, drainage and climatic conditions. The Groundwater conditions in State very widely. The depth of water varies from less than 5 mts to 20 mts in canal commands area whereas non- command area its varies from 7 mts to more than 100 mts. For evaluating trend of water level and long term fluctuation, year 1984 has been considered as base year. The trend had been analysed between pre-monsoon 1984 and pre-monsoon 2000 (Table 4.2).

As per studies, out of 32 districts, 26 districts show an average depletion of water level and 6 districts show marginal rise. However in the districts showing rise in water level there are certain pockets where depletion in water level has been noticed. The districts which have been severely affected by water level depletion are Ajmer, Alwar, Jaipur, Jalore, Jhunjhunu, Jodhpur, Nagaur, Pali and Sirohi.

Table 4.2: Change in Water Level between 1984 and 2000 in Different Districts of Rajasthan

Sl.No.	District	Average Pre-monsoon Water Level (m)		Drop/Rise in Water Level (m)
		Year 1984	Year 2000	
1.	Ajmer	7.82	13.41	(-) 5.59
2.	Alwar	11.08	16.20	(-) 5.12
3.	Banswara	6.35	6.351	(+) 0.001
4.	Baran	7.50	9.51	(-) 2.01
5.	Barmer	31.58	32.91	(-) 1.33
6.	Bharatpur	7.78	8.89	(-) 1.11
7.	Bhilwara	10.27	14.98	(-) 4.71
8.	Bikaner	65.34	62.87	(+) 2.47
9.	Bundi	9.23	12.53	(-) 3.30
10.	Chittorgarh	11.12	15.21	(-) 4.09
11.	Churu	42.88	42.84	(+) 0.04
12.	Dausa	11.06	13.24	(-) 2.18
13.	Dholpur	8.69	10.04	(-) 1.35
14.	Dungarpur	7.34	9.36	(-) 2.02
15.	Ganganagar	19.22	13.60	(+) 5.62
16.	Hanumangarh	22.51	18.09	(+) 4.42
17.	Jaipur	20.00	25.74	(-) 5.74
18.	Jaisalmer	45.29	45.06	(+) 0.23
19.	Jalore	13.18	21.18	(-) 8.00
20.	Jhalawar	8.13	9.35	(-) 1.22
21.	Jhunjhunu	28.32	34.54	(-) 6.22
22.	Jodhpur	30.23	36.83	(-) 6.60
23.	Karauli	11.04	14.47	(-) 3.43
24.	Kota	8.87	10.38	(-) 1.51
25.	Nagaur	28.39	34.89	(-) 6.50
26.	Pali	12.64	19.33	(-) 6.69
27.	Rajasmand	10.38	15.07	(-) 4.69
28.	Sawaimadhopur	9.58	11.32	(-) 1.74
29.	Sikar	30.91	35.68	(-) 4.77
30.	Sirohi	11.89	18.38	(-) 6.49
31.	Tonk	8.49	10.86	(-) 2.37
32.	Udaipur	9.25	11.92	(-) 2.67

Reasons of Groundwater Depletion

The prime reason for water table depletion is excessive withdrawal of groundwater to meet out ever growing demand for agriculture, domestic and industrial purposes. The sustainability of vagaries of rainfall had further fallen prey to the unplanned development. Thus result is shrinking of groundwater resources due to which wells are getting dried and chemical quality of groundwater is getting deteriorated. It is nothing but tragedy of the worst order that when both rainfall and groundwater area scarce commodity in the state and what ever and wherever available, it is under constant threat of being either polluted by industries or is being mis-managed in place and time in the name of 'so-called development' inflicting unbearable damage on their dependent fragile ecosystems.

During the severe conditions of 1984-88 and 1999-2000 the people at large faced unprecedented hardships of even procuring the drinking water supplies and were reeling under constant threat of infectious water-borne diseases, the rich-resourceful industries sector exploited the situation unscrupulously, using array of tankers for managing waters at the cost of not only agriculture but even the bare domestic drinking water needs of the helpess masses. Our undergroundwater table is rapidly falling and every next summer, our wells deepening or the older ones dry out-a persistent threat to agriculture which continues to be the prime occupation of the people throughout Rajasthan.

Change our relationship to water and that sustain it, all our wealth and knowledge will be meaningless. We are as dependent on freshwater for life as our ancient ancestors were. But many do not seen to be aware that this precious resource is disappearing. The clock is ticking but they do not know it.

India is well endowed with freshwater resources, misuse and random application of freshwater in agriculture, and Industrial purposes are causing concern. there wide variations of water availability rate in different regions of the country. India often suffers from periodic water shortage the classical example is the Cherapunji and Mawasynram in Meghalaya state which get the highest rainfall in the world but suffers from water shortage during rate winter and early summer. On the other hand Rajasthan suffers throughout the year from acute water scarcity.

Surface Water Resources

The annual surface run-off in the rivers in Rajasthan from rainfall is estimated as 19585 MCM equivalent to a mean runoff depth of 5-7 cm gives the surface runoff for different river basins in Rajasthan. Table 4.3 shows that about 15 per cent of the surface runoff is concentrated in the three perennial rivers of Banas, Mahi and Chambal flowing in the Southern and Eastern parts of the state.

Table 4.3: Surface Run-off in Different River Basins in Rajasthan

Sl.No.	*River Basin*	*Area km^2*	*Annual Flow (MCM)*	*Utilisable Flow (MCM)*
1.	West Banas	1999	800	90
2.	Mahi	16891	3570	3350
3.	Sabarmati	2797	890	840
4.	Banas	45543	4360	3100
5.	Chambal	31393	6130	2980
6.	Luni	34569	1630	530
7.	Sahibi	4532	280	150
8.	Banganga	6746	440	350
9.	Baran	3225	230	120
10.	Gambhir	4812	450	400
11.	Parvati	1968	200	180
12.	Sukri	945	230	50
13.	Others	10172	375	155
	Total		19585	12295

Source: Irrigation Department, Govt. of Rajasthan, Jaipur.

Owing to the limitations of topography, physiography and soils only about 12300 MCM of surface water can be utilised in addition to this about 15000 MCM is brought in by the river (Bhakra and Gang Canal) outside the Rajasthan. Thus as of today the total usable flow in Rajasthan is about 27300 MCM giving per capita amount of annual usable flow of the order of 680m^3. On the other hand, for India as a whole the per capita amount of annual usable flow is about 840 m^3. The surface water is disposed off in three ways. Part of it is stored in reservoirs, another part disappears as percolation from streams where groundwater table is just below the stream surface and rest finds its way to the sea. A part of water stored in reservoirs is lost though evaporation and seepage and the rest is utilized for various purposes mainly for irrigation.

Of the 1985 MCM of water that flows in the rivers annually at the present about 8650 MCM is stored various reservoirs and tanks. There is evaporation loss of the order of 30 per cent reservoirs and tanks. The percentage loss is greater in tanks because they are relatively shallow. The remaining flow of about 10935 MCM goes to the sea and adjoining States.

The surface runoff in different river basins is mentioned in above table. The State had number of rivers and streams with drainage area of over 500 km^2. The Aravallis from the water decide as a result the rivers of west Rajasthan and East Rajasthan do not have direct connection with one another.

The major rivers of East Rajasthan are the Banas, the Chambal, the West Banas, the Mahi and the Sabarmati. The Banas river rises in the Aravallis, about 65 km Northeast of Udaipur and after flowing for a distance of 480 km joint the river Chambal. The river Chambal with its tributaries (the Kalisindh, the Parvati, the Kun, the Mej, the Gambhir, the Morel and the Ken) after flowing in southwest of Etawah in Uttar Pradesh meets the Yamuna river. The river West Banas, Mahi and Sabarmati which are perenniag, flow south ward in the gulf of Cambay. The river Luni is the only major river in the West Rajasthan. It rises in the Aravallis near Ajmer and after a course of about 320 kms is lost in the Rann of Kutch. The Luni, the Mitri, the Sukri, the Guhiya, and the Bandi, the Jojri, the Sagi and the Jawai are the tributaries joining the river Luni from the left. the rivers of West Rajasthan are ephemeral in character and flow only when severe rainstorms occur over the river catchment. The inland rivers such as the Kantli, the Sota the Sahibi and the Banas flow in the Semi-Arid Zone. These inland rivers after flowing for some distance are lost in the deserts.

Water Availability from Outside the State

Though without any perennial whole river of its own, Rajasthan is fortunately located surrounded by states having surplus run-off this national resource *i.e.*, water. The details and share of Rajasthan's agreements under inter-state river systems with them is revealed in

Table 4.4, shows its share and actual utilization of the resource leaving a wide gap of 5.31 MAF which needs to be bridged at the earliest throught purposeful efforts.

Table 4.4: Rajasthan's Share Under Inter-State River Systems Surface Water Supply

Sources	*States*	*As per Agreement the Share of Rajasthan (in MAF)*	*Water Actually Utilised (in MAF)*
Gang canal	(Punjab)	1.11	1.11
Bhakra	(Punjab)	1.50	1.50
Gurgaon canal	(Haryana)	0.09	0.09
Ravi-Beas	(Punjab)	8.60	3.59*(Stage#I)
Narmada	(M.P.)	0.50	–
Bharat Feeder	(U.P.)	0.02	0.02
Chambal Kota Barrage	(M.P.)	1.60	1.60
Total		13.42	8.11

Out of 3.59 MAF from source, 3.27 MAF is for irrigation and 0.22 MAF is for drinking purposes. Under IGNP Stage II, 4.0 MAF water is allocated of which 3.35 MAF is for irrigation and 0.65 MAF is for drinking purposes.

Source: Government of Rajasthan, Report of Panel on Water Resources of Rajasthan, August 1989

Looking to our over-all environmental frame in general and rainfall behaviour in particular, the state should act vigilantantly in demanding its due share from the river systems of the neighbouring states in view of the new river valley projects emerging there; particularly along the courses of river Ravi-beas in Punjab, Ghaggar in Haryana. Yamuna-Ganga in U.P. and Narmada-Chambal etc. in Madhya Pradesh, etc.

Consumption Pattern of Water Resources

The growth in population and the expansion of economic activities have led to increasing demand of water fro diverse purposes. The main man use how ever is for irrigation which account for nearly 91 per cent of the water use and for drinking. Water which claims bulk of the remaining supplies.

Agriculture-irrigation combine has been the traditional maximum user of water, followed by domestic consumption needs of the people in both urban and rural areas. During the last 4 decades, industries, recreation-tourism union, sewerage needs in urban areas have started their respective claims with thrust and pull of resource fullness. Thus till recent times, while irrigation and drinking needs have accounted for nearly 99 per cent of the total available water in the state of which nearly 70 per cent came from surface water sources and the remaining 30 per cent from groundwater area.

Table 4.5: Summary of Non-Irrigation Water Demand Projection k, MM^3/Yr

Planning Stage	*Domestic*	*Industrial*	*Livestock*	*Other*	*Total*
1995	2002.0	47.2.	1069.0	167.0	3285.2
2005	2584.0	57.2	910.0	645.0	4196.2
2015	3176.0	71.4	1089.0	715.0	5051.4
2045	4724.0	137.7	1340.0	1867.0	8068.7

However, in near future growth of population, expansion of varied economic activities including industries, tourism and recreation, urban demands for better and higher living standard for all in addition to accelerated demand of drinking water in both rural and urban areas and irrigation for more assured agriculture will necessitate further thrusts for systematic study of water availability and/or augmentation project and schemes. Improvements in area under irrigation and double cropping might grow from the present 20-25 per cent of our total cropped area to even 40 or above per cent, putting extra-premium on scarce water resources in the state.

By the year 2015 A.D. our rural and urban population is projected around 5.2 and 2.6 crores respectively which will demand water around 4265 MM^3/yr respectively for only drinking purposes (for both human and livestock population-Report of state water Resources plan, GOR, Jaipur). The demand for additional quantity of water by industries, tourism and recreation and urban areas has already started mounting-up and progressively their shares in the total supply of water will inevitably shoot-up bringing in its wake damaging consequent misbalance in the fragile resource frame.

Management Issues

The above discussion has raised multi-faceted problem related with availability, fluctuations and variations of water resources in different areas and manifold competitive demands as part and parcel of developmental process and approach. Specific issues regarding hurdles in water management may be listed as:-first the inadequacy of the availability of water resource compared to the minimum basic needs of the population and this gap is sure to widen further under the pressure of higher rate of population growth second the spartio-seasonal variations of rains and water resources are most pronounced in Rajasthan and this variability and fluctuations are on recent records further increasing; third disappointingly and as a phased regular activity, specially in and around the rural areas (under one excuse or garb and name), the traditional methods of water conservation in practice, were systematically deserted, deteriorated and even destroyed-at places either obliterated or becoming even extinct; fourth the wasteful use of water in agriculture and irrigation sectors, by the urban population in a false-sense of higher level of living reflected in and measured through high per capita consumption of water when in their very neighbourhood the poorer sections of society including slum dwellers pine and precariously manage even their drinking water needs; and fifth the ever-raging, dominantly powerful and widening demands of industries and multiple urban activities (including tourism) flourish and expand, despite their wasteful and polluting behaviour pattern, at places at the cost of agriculture-cum-irrigation and even the basic facility of drinking water. This competition is to be observed in and around the growing urban-industrial-tourist centers of the state.

Planning Strategy for Sustainable Development

The sustainability of water can be reached only it we seek greater regional self sufficiency, not less. Building our economies on local watershed system is the only way to integrate sound environmental policies with people's productive capacities and to protect our water at the same time.

The values of unlimited growth and ever increasing economic globalization are in computable with the search for solutions to water scarcity. Economic globalization under mines local communities by allowing for easy mobility of capital and theft of local resources. In Rajasthan regardless of varied landform features, most of which suffer from the inadequly of water, as a base line approach strengthening their data of available and potential water resources but reliable at the same time, is essentially desired. However the few critical areas deserving immediate attention, considerable pondering and application of scientific research and analysis may be adopted following strategies to meet the challenge:

1. Comparative study of varied water management systems, planning policies and their implementation and target achievements in the state particularly in respect of irrigation and cropping patterns, impact of new technology and cultivators' responses to it. To investigate the extent to which the farmers have been able to accommodate and adjust the traditional know-how and modern technological innovations of water as such.
2. It is high time to undertake a scientific approach for delimitation of water zones in Rajasthan taking both the surface and groundwater resources availability into account on one hand and evaluation of the relative impact and utility of well, tank and canal irrigations in spatio-temporal perspectives on the other. An analysis of their problems with a view to improve their management, keeping conservation of water in view as one of the priority aspects, can go a long way in the purposeful utilization of our precious but limited water resources.
3. Roof top rain water harvesting programmes in major cities and towns and plan to implement groundwater augmentation schemes in rural areas.
4. Artificial recharge of groundwater by way of construction of different structures such as sub surface barriers, percolation tanks, Gabion structures, etc. identify critical areas and notify them. Revival of old traditional village pond tank system.
5. Change of cropping pattern (growing crops requiring less water) and use of improved irrigation methods should be encouraged.
6. To regular withdrawals from groundwater in over exploited and critical areas, ban on construction of new

structures, regulate indiscriminate boring by way of legislation etc.

7. There is an urgent need to once again examine performance of various irrigation systems, their relative economic gains and utility in an effort to search alternative usages including growing of plantations, grass animal farming and even fuel-forestry replacing the traditional cropping patterns for the greater and assured gains to the wider field of farming community. Similar assessment will prove beneficial if in a fresh bid, the entire gamut of canal water use in Rajasthan and management of tank irrigation in southern Rajasthan is re-examined and reviewed. It will be of certain gains if canals are made to operate at their targeted efficiency and the tanks in dis-use or lost are brought back to life to revitalize rural water scenario as a whole.
8. Time is quite ripe to take stock of present day utilization pattern of water resources in the state by agriculture, domestic and drinking water needs, industry, tourism-recreation etc. by examining their respective future shares and claims and competitiveness under projected demands. In this respect, people's responses to scarcity situation can provide some lessons for future management and planning insights, being purely indigenous and of practical gross-root experience.
9. Augmenting water supply by thorough and total utilization of there principal and major areas of (i) the available rainfall and river valley flows within the state, (ii) procuring water in all possible quantities from inter-state river agreements by remaining more vigilant about its share and (iii) exploring all channels and technical know-how for tapping even deeper groundwater resource (though its limitedness as an exhaustible stock resource character should never be lost-sight technique may help ease the situation, at least temporarily, by locating never and even deep groundwater reservoirs in more deficient areas, specially not touched by surface water sources.
10. All water management including its planning as well should aim at its conservation as a priority item,

particularly in Rajasthan. Like in energy, an unit saved is unit produced', in water also in Rajasthan in virtually all its parts (because by now no area of state has remained unvisited by drought, including Kota and Udaipur regions and now its Jaipur's turn for a much severe ensuing summer) conservation and saving for better utilization is an effort in its production cycle in physico-natural sphere *i.e.*, conservation of rain water and in human-economic-cultural sphere by regulation, management, judicious and scrupulous use of the supplied water with a view to conserve it also.

11. In a way conservation and efficiency of water-use has much to achieve in irrigation sphere, (by improved regulation of the main canal and its distributaries, their lining and entire aspect of operational techniques), in urban region (where the households-the biggest user culprits of being 'islands of opulence amid ocean of water scarcity' reflecting wide distributional preferential mismanagement within a unit), industrial and recreational segments. A more rational approach may include aspects like cost-effective pricing of water consumed by individual family, equity and justice in water supply and some kind of moral institutional control on private ownership of water through tube wells, etc. No less interesting is to evaluate the productive capacity of per unit of water utilized under various use-heads.
12. Last three decades in particular have witnessed acdrue environmental and ecological problems and hazards around industrial complexes related with their use of water. In this interdisciplinary field where technocrats, administrators, social-scientists and may other have to contribute collectively to maintain the fragile agrarian base in the state, diversify its economy involving shifts in the traditional structure without causing further irreversible damages to physico-human environmental balance. Desired safe-guards including treatment plants should invariably form part of the production process to avoid repetitions of environmental repercussions experienced earlier around industrial and mining areas of the state.

13. A provision should be made in the construction by-laws to the effect that if there is no sewerage in the area the owner will be obliged to let out dirty water of the house into the a pit 70-80 feet deep. This will prevent spread of diseases and help augment the underground water resources. Such a mandatory condition exists in Tamil Nadu.
14. In the rural areas, particularly in the desert areas, incentives should be provided to the people to enable them to construct underground tanks (Tankas) so that rain water can be collected and put to better use.
15. People should be taught the advantages of sprinkler and drip irrigation and should be encouraged to adopt these methods by providing them with incentive grants and liberal loans from banks.
16. Sustained efforts have to be made by the Govt. and various organizations through training, motivation camps, posters, wall paintings, hoardings, news papers, radio and T.V. etc. to make the water consumers aware of.

References

Agarwal, Anil and Narain, S., 1999. *Dying Wisdom: Rise Fall and Potential of India's Traditional Water Harvesting Systems*. Centre for Science and Environment, New Delhi.

Barlow, M. and Clarke, T., 2004. *Blue Gold: The Fight to Stop the Corporate Theft of the World's Water*. Leftword, New Delhi.

Black, Maggie, 2005. *The No-Nonsense Guide to Water*. Rawat Publications, Jaipur.

Gupta, N.L., 1960. Scope of landuse studies in Rajasthan. In: *Report of All Rajasthan Social Sciences Seminar*, Udaipur, pp. 17–29.

Gupta, N.L., 1973. Morphometric analysis of Udaipur basin. *Studies in Geography*, University of Rajasthan, Jaipur, 4(1972–73): 20–26.

Gupta, N.L., 1973. Rajasthan mein bar-bar sukha kuyon. *Lok Vigyan*, Udaipur, Special Issue on Famine year, 13(4–5): 22–29.

Gupta, N.L., 1980. Application of aerial photography in survey planning of watersheds for dryland agriculture. In: *Report of*

Dryland Technology under Integrated Rural Development Programme, Udaipur, pp. 123–141.

Gupta, N.L., 1980. *Water Resources at 2000 A.D.: A Case study of Udaipur Basic, Problems and Prospects*. UEG and Sewa mandir Publications, pp. 12–24.

Gupta, N.L. *et al.* (1992). *Journal of Water and Land Use Management*, Marudhara Acadamy, Jaipur.

Gurjar, R.K., 1987. *Irrigation Impact on Desert Ecology*. Scientific Publisher, Jodhpur.

Gurjar, R.K. and Gupta, N.L., 1993. *Integrated Water Use Management*. Rawat Publication, Jaipur.

Gurjar, R.K. and Shukla, Lakshmi, 1998. *Water Resources: Environment and the People*. Pointer Publishers, Jaipur.

Gurjar, R.K. and Jat, B.C., 1999. Watershed development programme: A scenario, a participatory issue. *Journal of Water and Land Use Management*, Marudhara Academy, Jaipur, 6(1).

Gurjar, R.K. and Jat, B.C., 2001. *Jal Prabandh Vigyan*. Pointer Publishers, Jaipur.

Jat, B.C., 1999. *Watershed Prioritization of Bandal Watershed: A Remote Sensing and GIS Approach*. Indian Institute of Remote Sensing, Dehradun.

Jat, B.C., 2000. *Jalgrahn Prabandhan*. Pointer Publishers, Jaipur.

Jat, B.C., 2002. Deepening water crisis. *Rajasthan Patrika* (Ed).

Jat, B.C., 2004. Climatic change and water crisis. Editorial in *Rajasthan Patrika*, Jaipur.

Jat, B.C., 2004. Rajasthan main jal prabandhan. In: *National Seminar on Environment and Sustainable Development*, Jaipur.

Jat, B.C., 2000. Application of remote sensing and GIS to water resources management. In: *National Seminar on Groundwater Management Strategies in Arid and Semi Arid Regions*, State Groundwater Department, Jaipur.

Kumar, Alok *et al.*, 2000. *Groundwater Resources of Rajasthan: Its Past, Present and Future Status and Need for its Management*.

Kumar, Amaresh, Kumar, Alok and Mathur, S.B., 2000. Trend of groundwater levels and rainfall pattern in Rajasthan (pre-monsoon 1984–pre-monsoon 2000).

Mahnot, S.C., Sharma, D.C., Mishra, A., Singh, P.K. and Roy, K.K., 2003. *Water Harvesting Management*, SDC/ICU, Jaipur.

Narwani, G.S., 2005. *Community Water Management*. Rawat Publications, Jaipur.

Singh, A.L. 1999. Planning for the optimal utilization of water resources for agricultural purposes: A case study of western Uttar Pradesh. In: *Water Resources Management*, (Ed.) A.F. Khan, B.R. Publishers, New Delhi, pp. 295–312.

Singh, A.L., 2003. Urban water supply and occurrence of diseases: A case study of the low income households of Aligarh City.

Singh, J., Singh, D.N. and Raju, K.N.P. (Eds.). *Water Crisis and Suitable Management*, T–B.A., Varanasi, pp. 250–65.

State Water Resources Plan, 1998. Irrigation Department, GOR, Jaipur.

Chapter 5

Groundwater Contamination and its Remediation Strategies: Towards the Challenge of Providing Safe Drinking Water

Raaz Maheshwari[1], *B.C. Jat*[2] *and Priyanka Singh*[3]

[1]*Head, Department of Chemistry, SBDTC, Lakshmangarh, Rajasthan*

[2]*Department of Geography, Govt. P.G. College, Neemkathana, Sikar, Rajasthan*

[3]*Research Scholar (Pharmaceutics), U.P. Technical University (SC-RKGIT), Ghaziabad*

Introduction

It is ironic that many areas of the world face critical shortage of safe drinking water on a planet whose 3/4th surface is covered with water. Consider this only 3 per cent of all water in the world is freshwater, the rest (97 per cent) being sea/salt water. Within this 3 per cent about 69 per cent is stored in glaciers, icecaps and snowy mountain ranges leaving only 1 per cent of total earth's water for falling our daily requirements. This readily available freshwater is mostly beneath the ground and only 0.3 per cent is in surface water bodies such as rivers and lakes.

Like water in streams and rivers, groundwater moves but at a very slow rate.

Groundwater flow from areas where precipitation percolates down to the water table, called recharge zones, to locations where it flews out of an aquifer and becomes surface water. Year 2003 was called the 'International Year of Freshwater' Speaking on the subject UN Secretary General Kofi Annan said " No single measure would do more to reduce disease and save lives in the Developing world than bringing safe water and adequate sanitation to all".

Per capita freshwater supplies are dipping alarmingly in poorer countries due to runaway population growth, deterioration of groundwater reservoirs, as well as primitive sanitation and waste treatment facilities[1] contaminated water kills approximately 40,000 people every day and is responsible for causing 80 per cent of the global health problems a wide variety of micro-organisms foul the water sources of devolving countries throughout the world spreading diseases such as diarrhoea, typhoid, cholera, dysentery and other water-borne infections that are often virulent in nature. Diarrhoeal diseases alone are responsible for taking the lives of more than six million children annually. In the last ten years, there has been an enormous increase in the number of fatalities from consumption of contaminated drinking water all over the world. What is specially galling is the fact that this uncalled for death and debilitation is occurring in areas where there is no major catastrophe or disaster such as earthquake tsunami, draught or famine. And when an official emergency actually does strike, the dispossessed populace has to take refuge under common shelters and eke out the use of meager water supplies available; drinking and washing from the same source, inevitably pollutes the water. This further boosts the water statistics concerning the number of people who have suffered or have lost their lives after drinking contaminated water.

The rural areas in the developing world consists of more than 70 per cent of the population. Water supplies is specially crucial for those dependent on agriculture. Here, the primary source of drinking water is often a shallow well (<10 feet deep) or a pond- usually badly polluted, since human and animal wastes are emptied directly into those water sources without proper treatment. Rivers and streams in other areas are often polluted as well, with toxic substances from industrial plants in addition to sewage pipe discharges. Waste material from industries are released carelessly into the water and may contain organic and inorganic material and that are hazardous to human health (2).

Subsequent to a natural calamity anywhere in the world, dearth of clean water emerges as one of the greatest health hazards because most of the water sources get polluted by a load of debris and waste material, as was seen in the tsunami-hit areas as well as in the Orleans after Hurricane Katrina struck. One of the primary task of the government and international relief organisations then becomes providing safe and clean water; otherwise disaster takes no time to invite diseases of many kinds-mostly related to the consumption of contaminated water.

About 200 million Indians are thirsting for safe and clean water, it is really shocking that India has only 4 per cent water resources, where as constitutes 16 per cent of total world population which is rising and likely to touch almost 25 per cent in 2050, the situation will worsen in time to come. the report " Safe Drinking Water" brought out by third World Academy of Sciences, reveals that people in most part of Asia and Africa don't have assess to safe drinking water and even in relatively advanced countries such as India state drinking water is not readily available particularly in rural areas.

It should not come as a surprise to any body that India ranks 120th among the group of 122 countries evaluated for water quality by the UN. This binding comes close on the heels of the discovery of more than the tolerable level of pesticides in the country's packaged drinking water. This indicates that freshwater on earth is finite and also unevenly distributed, This is when about 80 per cent of our body is composed of water and water is essential to most of our activities. Much of the natural supply of potable water is assessable faces stress from a growing world population which increases the basis demand of this natural resources, while reducing the supply further through biological and industrial contamination with established waste treatment or water treatment infrastructure often suffer from epidemic of water-borne diseases. In order to bridge the gap between the actual phase of emergency and the time when proper infra structure is reestablished in a disaster stricken area, a portable water purification dance would prove to be an ideal tool for providing clean drinking water and offering protection from water-borne diseases.

Groundwater Contamination

Inspite of understanding the importance of this natural resource and its limited availability man has been irresponsible towards protecting it from getting contaminated with our activities. We have allowed harmful and unhealthy toxicants to contaminate our groundwater resources.

Sewage is the most common source of groundwater pollution. Landfills, under ground storage tanks, disposal of untreated wastewater and other materials and disposal of untreated wastewater and various toxicants are various sources of contamination. Any chemical that is easily soluble and penetrate the soil is prime candidate for groundwater pollution. It is very difficult to cleanup contaminated groundwater, especially to the drinking water standards. However, given the seriousness of the issue extensive researches have been carried out and groundwater cleanup for remediation of contaminated soils. We know, practically, that installing and successfully running a water treatment plant is not a easy task, especially running of a treatment system for cleaning undefined contamination in gallons and gallons groundwater.

It is not easy to know contamination in groundwater. It might take weeks, months and sametimes years to investigate. That groundwater is contaminated. Contamination generally travel unnoticed in the subsurface untill found in a water supply well. Effect of the contamination depends on the concentration of the contaminants. Some contaminants cause serious health hazard ern with very low level. Therefore remediation becomes a necessity.

Type of Contaminant and Remediation Strategies

Generally remediation technologies means giving a remedy. The term in environmental parlance usually refers to the removal of pollution or contaminants from environmental media such as soil, groundwater sediment, or surface water for the general protection of human health and the environment. The goal of groundwater remediation is to restore the groundwater to its potential beneficial uses–in turn, protecting human health and the environment. At times soil remediation is done along with groundwater remediation as soil acts as the contaminant transfer media. To stop further percolation of contaminants to groundwater, it becomes necessary to cleanup contaminated soils. However here our focus is to discuss technologies used for removal of groundwater pollution.

It is important to understand the problem to provide the solution. The way standard treatment schemes have been developed overtime for water treatment depending upon quality of water available, similarly for remediation of contaminated groundwater, treatment solution have been evolved. As said earlier in India remediation is relatively new concept and its economic viability in Indian context is still to be ensured. However, this concept is not new in developed countries like us and extensive researches have been under taken to provide answers to complex contamination problems.

The United States Environmental Protection Agency (USEPA) maintains detailed information on its website on remediation technologies and their application. For application of remediation technology/ies, contaminants are generally separated into following contaminant groups:

- ☆ Non-halogenated volatile organic compounds (VOCS)
- ☆ Halogenated volatile compounds
- ☆ Non-halogenated semi-volatile organic compounds (SVOCS)
- ☆ Halogenated semi-volatile organic compounds
- ☆ Fuels, inorganics, radionuclides, explosives, pesticides, weedicides, fertilizers, fundamentally three primary strategies are used separately or in conjunction for remediation of contamination.
- ☆ Destruction or alteration of contaminants.
- ☆ Extraction or separation of contaminants from environmental media
- ☆ Immobilization of Contaminants

Treatment technologies capable of contaminant destruction by altering their chemical structure are thermal, biological and chemical treatment methods. The destruction technologies can be applied *in situ* or *ex situ* to contaminated media.

Treatment technologies commonly used for extraction and separation contaminants from separation contaminants from environmental media include groundwater treatment by either phase separation, carbon adsorption air stripping, ion exchange reverse osmosis, electro-dialysis, photochemical oxidation, electro-coagulations, phytoremediation, microbial biosorption, or same combination of these technologies.

Immobilization technologies include stabilization solidification and containment technologies, such as placement in a secure–landfill or construction a slurry walls no immobilization technology is permanently effective, so some type of maintenance is desired. Generally, no single technology can remediate an entire site. Several treatment technologies are usually combined at a single site to get desired results.

In India, concept of remediation is new and not many studies have been initiated to identify and tackle such problems. There have been incidents of groundwater contamination in India. Such as arsenic contamination in west Bengal, Ballia and Lakhimpur in UP, fluoride contamination in Rajasthan and 18 other states, and also high percentage of heavy metals *viz.* (V, Cd, Hg, Co, Mn....) in several states. The supreme court of India ordered the industries responsible for contamination to undertake remediation task. In India, lately, some studies have been initialed to assess groundwater contamination problem and accordingly identity technology suitable for undertaking remediation most of these studies are still in the investigation stage and on-ground remediation well plan in still to be finalized.

References

Kaur, H., 2006. The challenge of providing safe drinking water. In *Depth Everything About Water,* January–February, pp. 118–119.

Maheshwari, R., Arora, C. and Singh, P., 2006. Safe water: Best gift to future's mankind. In: *Proceedings of State Seminar on Excess Fluoride in Potable Water and its Associated Health Hazards,* Alwar August 5–6, pp. 66.

Maheshwari, R. and Raani, B., 2006. Towards conserving the elixir of life. In: *Proceedings of National Conference on Environmental Conservation (NCEC–2006),* Pilpovi, September 1–3, pp. 645–652.

Malik, S. and Agarwal, S., 2006. Groundwater: Contamination and its remediation. *Water Digest,* July–August, pp. 90–92.

Raani, B., 2006. How to save water is our source holds. In: *Proceeding National Conference on Save Water Save Earth,* Jaipur September–November 25–26, pp. 170–172.

Chapter 6

Food and Nutrition Security: The Challenges of Climate Change

Reeta Mishra[1]*, Y.P. Singh*[1]*, Y.D. Mishra*[2]*,*
B.P.S. Raghubanshi[1] *and Swati Singh*[1]
[1]*RVSKVV-Krishi Vigyan Kendra, Morena, M.P.*
[2]*Directorate of Extension Services, RVSKVV-Gwalior, M.P.*

There is growing realization of the immense challenge of feeding the growing human population this century. Among the many reasons for this is concern that climate change will affect agricultural productivity, livelihoods, and development prospects. Most obviously, climate change will affect agricultural production by altering temperature and precipitation patterns. Climate change has long been predicted to alter the intensity and distribution of rainfall.

Climate change will increase the risk of hunger and under nutrition over the next few decades and challenges the realization of the human rights for health and adequate food. Climate change will affect nutrition through different causal pathways that impact food security, sanitation, water and food safety, health, maternal and child health care practices and many socio-economic factors. With a likely change in the patterns of climate-related extreme events, such as heat waves, droughts, storms, heavy precipitation and floods and increased risks of disasters, vulnerable communities and households will suffer serious setbacks in terms of food and nutrition security. Climate change negatively affects food availability, conservation, access and utilization and exacerbates socio-economic risks and vulnerabilities. Climate change is expected to further reduce food productivity and to make production even more erratic in regions where agricultural productivity is already low. With local production declining and probable disruptions caused by climate hazards, income generating opportunities and purchasing power will decrease for vulnerable populations. At the same time, decreases in production could lead to price increases for staple crops of 25 to 150 per cent by 2060. In regards to this, the present paper deals with objectives of seeking the relationship between agriculture and food security, impact of climate change on food and nutrition security and interventions to improve food and nutrition security.

Food Security

It is defined by the FAO as "*a situation in which all people at all times have physical and economic access to sufficient, safe and nutritious food to meet their dietary needs and food preferences for an active and healthy life.*" Food security has three following levels:

1. National/Regional Food Security

It is defined as a satisfactory balance between food demand and food supply at reasonable prices.

2. Household Food Security

Households are defined as food secure if their entitlements or demand for food is greater than their needs. Poverty has been identified as the root cause for household food insecurity.

3. Individual Food Security

At the individual level, an individual is considered food secure if his or her food consumption is greater than need as defined by physiological requirement.

Nutrition Security

The concept of food security cannot be made holistic without nutritional security. In the last two decades, there has been a progressive decline in pulse consumption, especially among the poorer segments of the population. This is due to stagnant production and the rising cost of pulses. Post-harvest losses especially in vegetables and fruits are presently in the range of 20 to 30 per cent and contribute directly to higher costs and reduce availability of these commodities.

Indian diets are qualitatively and quantitatively deficient in micro-nutrients due to inadequate intake of vegetables, fruits, pulses and millets which leads to nutrition related problems. Major nutrition-related public health problems are listed below;

1. Chronic energy deficiency and under-nutrition
2. Micro-nutrient deficiencies
   - Anaemia due to iron and folate deficiency
   - Vitamin A deficiency
   - Iodine Deficiency Disorders
3. Chronic energy excess and obesity

In India, the prevalence of anaemia is high because of low dietary intake, poor iron and folic acid intake; poor bio-availability of iron in phytate fibre-rich Indian diet; and infection such as malaria, hook worm infestations. The intake of Vitamin A is significantly lower than the recommended dietary allowance in young children, dietary adolescent girls and pregnant women. In these vulnerable sub-groups multiple nutritional problems coexist including inadequate intake of energy as well as of micro-nutrients other than Vitamin A. Unlike other micro-nutrient deficiencies, iodine deficiency disorders are due to deficiency of iodine in water, soil and foodstuffs and affect all socio-economic groups living in defined geographic areas.

Relationship between Agriculture and Food Security

Agriculture is important for food security in two ways: it produces the food for people to eat; and it provides the primary source of livelihood. If agricultural production in low-income developing countries of Asia and Africa is adversely affected by climate change, the livelihoods of large numbers of the rural poor will be put at risk and their vulnerability to food insecurity increased. The food security implications of changes in agricultural production patterns and performance are of two kinds:

1. Impacts on the production of food will affect food supply at the global and local levels. Globally, higher yields in temperate regions could offset lower yields in tropical regions. However, in many low-income countries with limited financial capacity to trade and high dependence

on their own production to cover food requirements, it may not be possible to offset declines in local supply without increasing reliance on food aid.

2. Impacts on all forms of agricultural production will affect livelihoods and access to food. Other food system processes, such as food processing, distribution, acquisition, preparation and consumption, are as important for food security. More frequent and more intense extreme weather events (droughts, heat and cold waves, heavy storms, floods), rising sea levels and increasing irregularities in seasonal rainfall patterns (including flooding) are already having immediate impacts on not only food production, but also food distribution infrastructure, incidence of food emergencies, livelihood assets and human health in both rural and urban areas.

These will affect the suitability of land for different types of crops and pasture; the health and productivity of forests; the distribution, productivity and community composition of marine resources; the incidence and vectors of different types of pests and diseases; the biodiversity and ecosystem functioning of natural habitats; and the availability of good-quality water for crop, livestock and inland fish production. Arable land is likely to be lost owing to increased acidity (and associated salinity), groundwater depletion and sea-level rise. Food systems will be affected by internal and international migration, resource-based conflicts and civil unrest triggered by climate change. Climate change affects the fundamental requirements for health–clean air, safe drinking water, sufficient food and secure shelter. Many of the major killers such as diarrhoeal diseases, malnutrition, malaria and dengue are highly climate-sensitive and are expected to worsen as the climate changes.

Climate change induces various types of stresses on livelihoods, resulting in less food being produced or less money to buy food. Significantly negative impacts of climate change on food security.

Impacts of Climate Change on Food and Nutrition Security

Food security exists when all people in a community or other spatial unit, at all times, have physical and economic access to safe and nutritious food (and food preferences) that is sufficient to meet their dietary needs for an active and healthy life, and is obtained in a socially acceptable and ecologically sustainable manner. Climate change can affect food security because of the dependence of food systems on the mean and variability of weather.

The four dimensions of food security are food availability (*i.e.* production and trade), stability of food supplies, access to food, and food utilization. Food security is often used interchangeably with malnutrition. However, malnutrition indicates various forms of under nutrition that are caused by many factors, including dietary inadequacy, infections, and socio-cultural factors. Under nutrition includes stunting, wasting, and deficiencies of essential vitamins and minerals, as well as obesity or over-consumption of specific nutrients.

Climate change will affect all four dimensions of food security. Agricultural output in developing countries is expected to decline by 10-20 per cent by 2080. Globally, the potential for food production is projected to grow with increases in local average temperature over a range of 1-3°C, but above this it is projected to decrease. In seasonally dry and tropical regions, even slight warming (1-2°C) reduces yield. Changes in the patterns of extreme weather events will affect the stability of, as well as access to, food supplies. Increasing frequency of crop loss due to these extreme events may overcome positive effects of moderate temperature increases. For forests, elevated risks of fires, insect outbreaks and wind damage are projected. Food insecurity and loss of livelihoods would be further exacerbated by the loss of cultivated land. Climate-related animal and plant pests and diseases will reduce food availability, influence production system stability and reduce food access through reduction of income from animal production, and lower crop yields, as well as increased costs of control. Climate change may affect health outcomes and food utilization with additional malnutrition consequences. For example, populations in water-scarce regions are likely to face decreased water availability. Flooding, increased precipitation and higher temperatures are likely to increase the incidence of infectious and diarrhoeal diseases. Climate change is projected to increase the burden of diarrhoeal diseases in low-income regions by approximately 2-5 per cent by 2020. A rise in the food bill for households that are net buyers of food may lead to the substitution of starchy staples for micro-nutrient-rich animal source foods, legumes, processed foods, fruits and vegetables and to a reduction in the average number of meals and the amount of food consumed and therefore to reduced micro-nutrient intakes among poor people. Extremely poor people will experience decreased calorie consumption.

Impact of Climate Change on Under-Nutrition

Maternal and child under-nutrition is the underlying cause of 3.5 million deaths each year and 35 per cent of the disease burden in children younger than 5 years. The recent food and economic crises combine with the growing threat and negative impacts of climate change. Climate change is happening now and it represents a major threat for the coming decades. Climate change impacts both natural and human systems from the global to the local levels. Inadequate dietary intake and disease are the immediate causes of under-nutrition. Inadequate food consumption heightens vulnerability to infectious diseases, which, in turn, can prevent the body from absorbing adequate food. These immediate causes stem from underlying causes characterized by insufficient access to safe and appropriate food, inadequate maternal, infant and young child care practices, poor access to quality water, sanitation and health services.

Impacts on Food and Water Security and Nutrition

Climate change is likely to exacerbate the declining reliability of irrigation water supplies leading to increased competition for water. Water scarcity may lead to multiple adverse health outcomes, including waterborne diseases, exposure to chemicals, vector-borne diseases associated with water-storage systems and malnutrition. Drought and water scarcity can have negative effects on nutrition and the HIV/AIDS epidemic may further amplify these effects. Children in poor rural and urban slum areas are at high risk of diarrhoeal disease mortality and morbidity.

Interventions to Improve Food and Nutrition Security

Hundreds of millions of people are currently at a degree of risk from the impact of global climate change. This number is likely to increase in the future, not only because of the increasing risk and multiple threats of climate impacts, but also due to the largest consequences and negative impact on livelihood and nutrition security. Climate change amplifies the risk of under-nutrition, given that disaster risk, seasonal stresses and livelihood insecurity will be magnified. The ultimate causes of food insecurity and under-nutrition are social, economic, cultural and political. Therefore, it is essential that efforts to achieve food security and good nutrition address these causes. The following steps are required:

- Increase investments in human resources, sustainable food production systems and rural development;
- Implement policies to improve physical and economic access by all to sufficient, nutritionally adequate and safe food and its effective utilization;
- Focus on participatory and sustainable agriculture;
- Use a "nutrition lens" to direct multi-sectoral actions to improve household food security; improve food quality and safety; prevent, control and manage infectious diseases and micro-nutrient deficiencies; promote appropriate diets, including breastfeeding and healthy lifestyles; provide care for the vulnerable, including people living with HIV/AIDS; introduce productive safety nets; and provide direct assistance.

The production and availability of food grains, among other things, is an important indicator of the food security situation of a country. Since agriculture growth is limited, imports can help improve the country's supply situation for a short term. But for the long term, the country, the country will need to focus on productivity enhancement, through public investment in irrigation, research and efficient use of water, plant nutrition and other inputs. At the national level, food security is ensured by increasing the production of food grains. The household and individual food security is ensured by increasing their entitlements in terms of their purchasing power.

Farming community should be motivated towards the use of new technology for increasing production and productivity of crops. IT tools makes it possible to disseminate the right information to the right people at right time and in right form. The performance of agriculture, especially food grains production, is therefore, extremely crucial to make a nation food secure.

Dietary diversification is the most appropriate and sustainable option ensuring adequate intakes of all micro-nutrients and photochemical. Availability, affordability, access and awareness are some of the major determinants of sustained dietary diversification in families and communities. Dietary diversification can be made possible through community effort through increased production of micro-nutrient dense foods and reduced wastage through appropriate processing. It can be linked to income generation, particularly for the rural women. The problem of under nutrition should be minimized by ensuring appropriate infant feeding practices (universal colostrum feeding, exclusive breast feeding up to six months, introduction of semisolids at six months); promoting appropriate intra-family distribution of food based on requirements; dietary diversification to meet the nutritional needs of the family

Millets are rich in minerals and micro-nutrients and hence increased consumption will improve the intake of these vital nutrients by the poor. Consumption of adequate quantities of vegetables, especially green leafy ones, is essential for meeting the dietary requirement of vital micronutrients. Horticultural products provide higher yields per hectare and sell at higher prices. The processing, storage and transportation of horticultural products should be done in a manner so that there is no glut and distress sales will make their production economically attractive to farmers and improve availability to the consumers. Innovative local efforts can go a long way in improving nutrition security especially for the poorer segments of the population living in vulnerable areas.

So the strategy to overcome the problem of food and nutrition insecurity includes adequate availability of foodstuff by:

- ✰ Ensuring production of cereals, pulses and seasonal vegetables to meet the nutritional needs.
- ✰ Making them available throughout the year at affordable cost through reduction in post-harvest losses and appropriate processing.
- ✰ Improving people's purchasing power through appropriate programmes.

Thus food security includes availability and access to nutritionally adequate food which includes proteins, minerals and vitamins in addition to carbohydrates. In this regards, the role of fruits and vegetables becomes crucial because of their nutritional value. Green leafy vegetables are a rich and cheap source of all micronutrients as well as health promoting phyto-chemicals. Hence, investment in horticultural crops is likely to prove an important milestone toward ensuring food security. It is evident that women are a major part of solution to curing the problem of food and nutrition insecurity. With more income, families have more control over their life styles and provide better food for their nutritional needs. Thus nutrition education should be given to women and their family members to promote dietary diversification to achieve a balanced intake of all micro-nutrients.

Climate change directly affects food and nutrition security, undermining current efforts to address under-nutrition, one of the most serious but least addressed socioeconomic and health problems. Unless we urgently do some efforts, it will not be possible to reduce hunger and under-nutrition under a changing climate.

References

Beddington, J., 2010. Food security: Contributions from science to a new and greener revolution. *Philosophical Transactions of the Royal Society B*, 365: 61–71.

Black, R.E., Allen, L.H., Bhutta, Z.A., Caulfield, L.E., de Onis, M., Ezzati, M., Mathers, C. and Rivera, J., 2008. Maternal and child under-nutrition: Global and regional exposures and health consequences. *The Lancet*, 371: 243–260.

Bloem, M.W., Semba, R.D., Kraemer, K. and Gandolfo, Castel, 2010. Workshop: An introduction to the impact of climate change, the economic crisis, and the increase in the food prices on malnutrition. *Journal of Nutrition*, 140: 132S–135S.

FAO, 2009. *The State of Food Insecurity in the World*. FAO, Rome.

Godfray, H.C.J., Beddington, J.R. *et al.*, 2010. Food security: The challenge of feeding 9 billion people. *Science*, 327: 812–818.

Gopalan *et al.*, 2004. *Nutrient value of Indian Foods*. National Institute of Nutrition (ICMR), Hyderabad.

Milly P.C.D., Wetherald, R.T., Dunne, K.A. and Delworth, T.L., 2002. Increasing risk of great floods in a changing climate. *Nature*, 415: 514–517.

Schmidhuber, L. and Tubiello, F.N., 2007. Global food security under climate change. In: *Proceedings of the National Academy of Science* (*USA*). 104: 19703–19708.

Zhang, X., Zwiers, F.W., Hegerl, G.C., Lambert, F.H., Gillett, N.P., Solomon, S. *et al.*, 2007. Detection of human influence on twentieth–century precipitation trends. *Nature*, 448: 461–465.

Chapter 7

Effect of Climate Change on Indian Agriculture

Y.D. Mishra[1], *Reeta Mishra*[2] *and Swati Singh*[1]
[1]*Scientist, Directorate of Extension Services, RVSKVV-Gwalior, M.P.*
[2]*Scientist, RVSKVV-Krishi Vigyan Kendra, Morena, M.P.*

Climate change can manifest itself in gradual changes in temperature, precipitation and a rise in sea level, resulting in changes in frequency, intensity and duration of extreme events. Global warming means earth getting warmer and resulting in an ecological imbalance. Climate change is impacting agriculture in countries like India. India is a large country with 15 agro-climatic zones, with diverse seasons, crops and farming systems. For a majority of people in India, agriculture is the main source of livelihood. Agriculture is the most vulnerable sector to climate change as it is inherently sensitive to climate variability and climate change will leave its impacts on Indian agriculture in various direct and indirect ways. This obviously means an impact on the lives and livelihoods of millions of Indians. Climate change and agriculture are interrelated processes, both of which take place on a global scale. Global warming is projected to have significant impacts on conditions affecting agriculture, including temperature, carbon dioxide, glacial run-off, precipitation and the interaction of these elements. These conditions determine the carrying capacity of the biosphere to produce enough food for the human population and domesticated animals. The overall effect of climate change on agriculture will depend on the balance of these effects. Assessment of the effects of global climate changes on agriculture might help to properly anticipate and adapt farming to maximize agricultural production.

About two-thirds of the sown area in the country is drought-prone and around 40 million hectares is flood-prone. The poorest people are likely to be hardest hit by the impacts of climate variability and change because they rely heavily on climate-sensitive sectors such as rainfed agriculture and fisheries. They also tend to be located geographically in more exposed or marginal areas, such as flood plains or nutrient-poor soils. The poor also are less able to respond due to limited human, institutional and financial capacity and have very limited ability to cope with climate impacts and to adapt to a changing hazard burden. In this regard, this paper deals with the objectives of studying different climate factors affecting agriculture productivity, impact of climate change on agriculture, and adaption of agriculture to climate change.

Climate Factors

There are several factors which directly connect climate change and agricultural productivity. These are as follows:

- Average temperature increase
- Change in rainfall amount and patterns
- Rising atmospheric concentrations of CO_2
- Pollution levels such as tropospheric ozone
- Change in climatic variability and extreme events

Average Temperature Increase

An increase in average temperature can 1) lengthen the growing season in regions with a relatively cool spring and fall; 2) adversely affect crops in regions where summer heat already limits production; 3) increase soil evaporation rates, and 4) increase the chances of severe droughts.

Change in Rainfall Amount and Patterns

Changes in rainfall can affect soil erosion rates and soil moisture, both of which are important for crop yields. The IPCC predicts that precipitation will increase in high latitudes, and decrease in most subtropical land regions–some by as much as about 20 per cent. While regional precipitation will vary the number of extreme precipitation events is predicted to increase.

Rising Atmospheric Concentrations of CO_2

Increasing atmospheric CO_2 levels, driven by emissions from human activities, can act as a fertilizer and enhance the growth of some crops such as wheat, rice and soybeans. CO_2 can be one of a number of limiting factors that, when increased, can enhance crop growth. Other limiting factors include water and nutrient availability. While it is expected that CO_2 fertilization will have a positive impact on some crops, other aspects of climate change (*e.g.*, temperature and precipitation changes) may temper any beneficial CO_2 fertilization effect.

Pollution Levels such as Tropospheric Ozone

Higher levels of ground level ozone limit the growth of crops. Since ozone levels in the lower atmosphere are shaped by both emissions and temperature, climate change will most likely increase ozone concentrations. Such changes may offset any beneficial yield effects that result from elevated CO_2 levels.

Change in Climatic Variability and Extreme Events

Changes in the frequency and severity of heat waves, drought, floods and hurricanes, remain a key uncertainty in future climate change. Such changes are anticipated by global climate models, but regional changes and the potential effects on agriculture are more difficult to forecast.

Impact of Climate Change on Agriculture

In the long run, the climatic change could affect agriculture in several ways:

- *Productivity*, in terms of quantity and quality of crops.
- *Agricultural practices*, through changes of water use (irrigation) and agricultural inputs such as herbicides, insecticides and fertilizers.
- *Environmental effects*, in particular in relation of frequency and intensity of soil drainage (leading to nitrogen leaching), soil erosion, reduction of crop diversity.

- *Rural space*, through the loss and gain of cultivated lands, land speculation, land renunciation, and hydraulic amenities.
- *Adaptation*, organisms may become more or less competitive, as well as humans may develop urgency to develop more competitive organisms, such as flood resistant or salt resistant varieties of rice.

The impact of climate change on agriculture could result in problems with food security and may threaten the livelihood activities upon which much of the population depends. Climate change can affect crop yields (both positively and negatively), as well as the types of crops that can be grown in certain areas, by impacting agricultural inputs such as water for irrigation, amounts of solar radiation that affect plant growth, as well as the prevalence of pests. An increase of 2°C in temperature could decrease the rice yield by about 0.75 ton/ha in the high yield areas; and a 0.5°C increase in winter temperature would reduce wheat yield by 0.45 ton/ha. Major impacts of climate change will be on rain fed crops (other than rice and wheat), which account for nearly 60 per cent of cropland area. In India poorest farmers practice rainfed agriculture.

It is projected that due to climate change, *kharif* rainfall is going to increase and this might be positive for kharif crops. Further, for *kharif* crops, a one-degree rise in temperature may not have big implications for productivity. However, temperature rise in *rabi* season will impact production of wheat, a critical food-grain crop. The surface air temperatures will increase by 2 to 4°C by 2070-2100. As mentioned earlier, the *rabi* crop will be impacted seriously and every 1°C increase in temperature reduces wheat production by 4-5 million tons. This loss can be reduced to 1-2 million tons only if farmers change to timely planting. Increased climatic extremes like droughts and floods are likely to increase production variability. Productivity of most cereals would decrease due to increase in temperature and decrease in water availability, especially in Indo-Gangetic plains. The loss in crop production is projected at 10-40 per cent by 2100, depending upon the modeling technique applied.

Adaption of Agriculture to Climate Change

Sustainable Agriculture

Sustainable Agriculture can be defined as an integrated farming system which is based on locally adapted cropping patterns and local resource (natural resources and natural processes) use based on local knowledge, skills and innovations.

The capacity of a farming system to adapt to changing climate and weather conditions is based on its natural resource endowment and associated economic, social, cultural and conditions. The viability of these elements also constitutes the basis for sustainable agriculture, understood as agricultural production that: ensures adequacy of food production; does not harm the resource base; is economically viable; and enhances quality of life. Many climate and weather risk management strategies fit squarely into sustainable agriculture practices and can, therefore, be promoted with several of the programs and policies targeting environmentally responsible production.

Strategies to be Adopted

1. Changes cropping patterns and cropping systems to suit the local resource and weather conditions. Multiple/ mixed cropping, intercropping systems with legume components etc.
2. Ecological farming practices which can maximize the local resource use. Many of these practices are based on indigenous knowledge and focus on building soil biological productivity. Non Pesticidal Management, Organic Soil Management, Community Seed Banks, System of Rice Intensification, soil moisture management etc have already proven to be useful.
3. Locally adopted crop varieties specially in saline and flood prone areas, drought prone areas, making suitable selections adopting Participatory Plant Breeding and Participatory Varietal Selection.
4. Developing suitable farming systems integrating agriculture, horticulture and livestock.

Farmers' Institutions

Organized communities have proven to be more effective in planning and managing their resources and livelihoods, lobbying for a policy change and securing their entitlements. Appropriate institutional systems for each of the purposes need be established.

Strategy to be Adopted

1. The farmers would be organized into common interest groups federated into producer collectives. Existing institutions like Women SHGs etc would be used to initially anchoring the program.
2. These institutions would take the roles of planning, mobilizing resources, organizing production, and take up post-harvest management and marketing activities.
3. The producer collectives will improve the collective bargaining power of the farmers; will internalize market activities like bulking, primary and secondary processing which improve the village economy.

Food and Livelihood Security

Shift to sustainable agriculture is often seen as a compromise on food security. This is mainly because food is understood as only wheat and rice, few pulses, oilseeds and vegetables. The food basket can be increased if we can expand the scope to include millets, coarse cereals, dry land fruits, uncultivated greens etc which can also bring in nutrition security. Data from National Centre for Organic Farming (NCOF), ICRISAT and CMSA have proven that crop productivity can also be maintained with organic/ecological farming. Going beyond the current food security systems like PDS and mid-day meal schemes, systems need to be established to improve livelihood security in terms of sustaining food production in the village, improving income generation opportunities to the small farmers and agriculture labor is important in the rural areas especially in rainfed regions. The frequent monsoon failures results in droughts and support systems needs to be build to help the farm families and livestock to tide over.

1. Building house hold food security systems by adopting suitable cropping patterns

2. Village level management systems for alternative models like grain banks would be appropriate and attempted.
3. Suitable off-farm and non-farm employment opportunities would be identified and promoted.

Financial Support Systems

Currently all the financial support systems to agriculture are given only for external inputs. We need to create proper support systems for farm internalized inputs, community based infrastructure, knowledge and skill building and sharing etc.

1. Direct Subsidies to farmers rather than input subsidies
2. Integrating NREGA with sustainable agriculture so that each farmer gets 100 labour days for farming can provide ample scope in this direction.
3. Explore tools like Direct Income Support which exist in many developed/developing countries need to be explored as decent living income cannot be explored only through pricing mechanism.

Partnerships

At the district level we need build partnerships between various governmental and non governmental agencies to implement the program. At the national level we need to build an alliance of Public sector research organizations, extension agencies, departments dealing with rural livelihoods and NGOs which are working on sustainable agriculture/organic/natural/ecological farming.

Thus there are following five steps for adaption of agriculture to climate change:

1. *Assist farmers in coping with current climatic risks*
   - ✰ Weather services, agro-advisories, insurance, community banks for seed and fodder
2. *Intensify food production systems*
   - ✰ Technology and input delivery systems, market links
3. *Improve land and water management*
   - ✰ Technologies for resource conservation and use efficiency

4. *Enable policies and regional co-operation*
   - ☆ Incentives to farmers for resource conservation and use efficiency, pricing of resources, credit for transition to adaptation technologies
5. *Strengthen research for enhancing adaptive capacity*
   - ☆ Varieties, resource conservation technologies, pest surveillance for improved assessments: mechanism for collection and dissemination of weather, soil, water and agricultural data

Conclusion

Climate change is likely to reduce yields of most crops in long-term. In short-term effects may be small. Increased climatic variability could, however, cause significant fluctuations in production even in short-term.

In long-term better adapted varieties are needed to adapt to multiple stresses linked with climate change. In short-term, several options relating to technology transfer, its adoption, appropriate policies, and land and water management are available which can help minimize negative impacts. Costs of adaptation are less understood but likely to be high; costs of inaction could be even higher. Adaptation practices take time to become effective.

Climate change is already a reality for Indian farmers and that conventional models of agricultural research and extension will fail to address the need of the hour unless some fundamental recasting is done. The immediate need for interventionist action precludes traditional models of research and support systems and requires alternative but urgent programmatic interventions, led by farmers' institutions and their local resources, knowledge and innovations. Sustainable agriculture, on the other hand, holds immense mitigation and adaptation potential, specifically in the context of climate change even as it improves rural livelihoods and addresses the ecological crisis in Indian farming (genetic erosion, land degradation, water depletion and contamination etc.).

References

Climate Change–only one cause among many for Darfur conflict (2007). IRIN. 28 June 2007. Retrieved 2008–01–22.

Parthasarathy, B. and Pant, G.B., 1985. Seasonal relationship between Indian summer monsoon rainfall and the southern Oscillation. *International J. Climatol.*, 5: 369–378.

Shukla, P.R., Sharma, S.K. and Ramana, Venkata P., 2002. *Climate Change and India: Issues, Concerns and Opportunities.* Tata McGraw-Hill Publishing Company Limited, New Delhi, pp. 317.

Chapter 8

Effect of Initial Soil Moisture and Nutrient Management on the Yield of Mustard

Y.M. Kool[1], H.S. Bhadoria[2] and A. Singh[2]*

[1]Dean, Faculty of Agriculture, RVSKVV, Gwalior – 474 002, M.P.

[2]RVSKVV, Gwalior – 474 002, Madhya Pradesh

ABSTRACT

Problems and inadequacies such as shortage of water in the system, excessive system losses and chronic shortage towards tail of the canal system are visible. Soil was treated through organic manure, tillage and moisture conservation during *Rabi* including provision of life saving irrigation. The mustard crop was higher water use efficiency. The water use efficiency (WUE) was more in (I_1) than (I_0) treatment. The net returns of mustard, Rupees 24395 were recorded for mustard. Similarly, the lower returns were obtained when the initial moisture was 11-12 per cent. The response to initial moisture availability was more prominent in mustard crop. Since, at 19-20 per cent initial moisture, the returns were Rs. 24395 while at 11-12 per cent moisture returns obtained were 15470 during the experiments. It is revealed that after withdrawal of the monsoon mustard should be sown as early as possible to utilize

* Corresponding author: E-mail: koolym@yahoo.co.in

available moisture in the soil for sustainable agriculture. Results indicated higher water use efficiency (129 kg cm^{-1} ha^{-1}) in mustard crop in Chambal command area.

Keywords: *Initial soil moisture, Nutrient, Yield and Sustainable agriculture.*

Introduction

In low rainfall areas, twin factors limiting the crop production are moisture deficiency and low fertility. Spectuclar increase in crop yields has been obtained by adequate fertilization in low rainfall areas (rainfed) of the country (Nagre, 1981). Initial soil moisture regime at sowing time plays a key role in germination development of root system, vegetative growth as well as food production. The optimum plant growth, quality and higher yield depends upon well developed root system. The restricted root growth resulted in delayed sowing which reduced uptake of nutrients and ultimate poor plant growth. Plant root growth of mustard was severely affected by inadequate moisture at the time of sowing, mechanical impedance and sub-optimal, soil temperatures. One of the ways is to use the nutrients in an integrated manner from all possible resources and maximizing the utiliation of applied nutrient by crops (Acharya and Bandopadhyay, 2002). Bhushan *et* al. (1977) reported that early sowing in proper moisture condition gave the highest crop yield in *Brassica compestris* variety toria in silty loam soil. Verma *et* al. (2005) found favourable effect of soil moisture conservation on the yield of pea at Solan (H.P). Therefore, attempt has been made with the objective to find out the optimum soil moisture condition for sowing of mustard (The major *Rabi* crops of the region) and the research findings have the practical utility for the farmers in low rainfall situation of Chambal region of M.P.

Material and Methods

The experimental area is located at the JNKVV, ZARS, Morena. The study area lies between 25° 15′ N–latitude and 70° 30′–longitude with an altitude 174.31 m. The Details of different treatments and initial soil moisture regimes are presented in Table 8.1. Nutrient application kg ha^{-1} was for Mustard (80N + 40 P_2O_5 + 20 K_2O).

Table 8.1: Presence of Initial Moisture Regimes Under Different Treatments (Initial moisture content) at the Time of Sowing

Treatments	*Initial Moisture (First Year) Per cent*	*Initial Moisture (Second Year) Per cent*	*Irrigation(I_0) and Effective Rainfall (cm) (First Year)*		*Irrigation(I_1) and Effective Rainfall (cm) (Second Year)*		*Irrigation Applied (Days after Sowing)*
T_1	19-20	19-20	8.0	1.05	15	1.05	45 DAS
T_2	17-18	17-18	8.0	1.05	15	1.05	45 DAS
T_3	15-16	15-16	8.0	1.05	15	1.05	45 DAS
T_4	13-14	13-14	8.0	1.05	15	1.05	45 DAS
T_5	11-12	11-12	8.0	1.05	15	1.05	45 DAS

*Total water (Irrigation applied + effective rainfall)

Mean of 4 replications biometrical parameters *viz.* (1) Plant height (cm) (2) Number of branches per plant (3) Number of pods per plant and (4) Weight of mustard (JM-1) were recorded at different stages.

Result and Discussion

The average moisture content was calculated from 0-15, 15-30, 30-60 and 60-100. The presences of initial moisture regimes under different treatments are presented in Table 8.2 for two consecutive years.

Table 8.2: Influence of Post Sowing Irrigation on Net Return of Mustard Yield

Treatments	Net Return Rs ha^{-1}	Mustard Yield kg ha^{-1}	WUE kg cm^{-1} ha^{-1}	Nutrient Use kg ha^{-1}		
				N	P	K
$T_1 I_1$	24395	2135	129	26.68	53.37	106.7
$T_2 I_1$	23120	2060	124	25.75	51.50	103.0
$T_3 I_1$	21080	1940	117	24.25	48.50	97.00
$T_4 I_1$	18275	1775	107	22.18	44.37	88.75
$T_5 I_1$	15470	1610	97	20.12	40.25	80.50
Mean	20468	1904	115	23.80	47.60	95.20

The mustard crop recorded higher water use efficiency when pre-sowing irrigation was applied (I_0) in all the treatments. The net returns of mustard, Rupees 24395 was obtained from $T_1 I_1$ for mustard. Similarly, the lower returns were obtained when the initial moisture was 11-12 per cent Rupees 15470 from ($T_5 I_1$).

Conclusion

The response to initial moisture availability was more prominent in mustard crop. Since, at 19-20 per cent initial moisture, the returns were Rs. 24395 while at 11-12 per cent moisture returns obtained were 15470 during the experiments. It is revealed that after withdrawal of the monsoon mustard should be sown as early as possible to utilize available moisture in the soil for sustainable agriculture.

References

Acharya, C.L. and Bandopadhyay, K.K., 2002. Effective input management for sustainable production. *Indian Farming,* 52(8): 42–49.

Bhushan, L.S., Varade, S.B. and Gupta, C.P., 1977. Influence of clod size on soil temperature, moisture tension and seedling emergence of field crops. *Journal of Agricultural Engineering,* 11: 20–24.

Nagore, K.T., 1981. Response of *kharif* sorghum varieties to varying levels of NPK under rainfed conditions. *Indian J. Agric. Resources,* 15(3): 166–177.

Verma, M.L., Sharma, R.K. and Bhandari, A.R., 2005. Effect of soil water conservation practices on root growth, yield, nutrient removal and quality of pea and tomato crops. *Journal of Indian Society of Soil Science,* 53(4): 602–607.

Chapter 9

The Anthropogenic Costs of Climate Change

P.B. Reddy[1]* and Sateesh Bino[2]

[1]*Department of Zoology, Govt. College, Nagda, Ujjain, M.P.*

[2]*Additional Superindent of Police,*
Department of Police, Morena, M.P.

Global warming is occurring and human-driven emissions of carbon dioxide and other greenhouse gases, as well as land-use change, are primarily responsible. Given current trends, temperature extremes, heat waves and heavy rains are expected to continue to escalate in both frequency and intensity, and the earth's temperature and seas will continue to rise. These conclusions lie at the heart of a 2007 report issued by the Intergovernmental Panel on Climate Change (IPCC), the world's foremost scientific body for the study of climate change.

This report aims at filling a gap in both the general public's understanding of climate change and in the senior policy-makers' toolbox. To date, the human impact of climate change has been a rather neglected area of research. Indeed, this review attempts to set out the detrimental effects people already suffer today due to climate change, as well as the far greater impacts it will have on the lives of

* Corresponding author: E-mail: reddysir@yahoo.co.in

the next generation. As such, the report is an attempt to fill an important gap in our collective knowledge and represents a plausible narrative of the human impact of ongoing climate change.

The Anatomy of a Silent Crisis

Science is now unequivocal as to the reality of climate change. Human activities, in particular emissions of greenhouse gases like carbon dioxide are recognized as its principle cause. This report clearly shows that climate change is already causing widespread devastation and suffering around the planet today. Furthermore, even if the international community is able to contain climate change, over the next decades human society must prepare for more severe climate change and more dangerous human impacts.

This report documents the full impact of climate change on human society worldwide today. It covers in specific detail the most critical areas of the global impact of climate change, namely on food, health, poverty, water, human displacement, and security. The third section of this report highlights the massive socio-economic implications of those impacts, in particular, that the worst affected are the world's poorest groups, who cannot be held responsible for the problem. The final section examines how sustainable development and the Millennium Development Goals are in serious danger, the pressures this will exert on humanitarian assistance, and the great need to integrate efforts in adapting to climate change.

Based on verified scientific information, established models, and, where needed, on the best available estimates, this report represents the most plausible narrative of the human impact of climate change. It reports in a comprehensive manner the adverse effects people already suffer today due to climate change within a single volume, encompassing the full spectrum of the most important impacts evidenced to date. The findings of the report indicate that every year climate change leaves over 300,000 people dead, 325 million people seriously affected and economic losses of US$125 billion. Four billion people are vulnerable, and 500 million people are at extreme risk. These figures represent averages based on projected trends over many years and carry a significant margin of error. The real numbers could be lower or higher. The different figures are each explained in more detail and in context in the relevant sections of the report. Detailed information describing how these figures have been calculated is also included in the respective sections and in the end matter of the report.

The alarming figures may prove too conservative. Weather-related disasters alone cause significant economic losses. Over the past five years this toll has gone as high as $230 billion, with several years around $100 billion and a single year around $50 billion. Such disasters have increased in frequency and severity over the past 30 years in part due to climate change. Over and above these cost are impacts on health, water supply and other shocks not taken into account. Some would say that the worst years are not representative and they may not be. But scientists expect that years like these will be repeated more often in the near future.

Climate Change through the Human Lens

Climate change already has a severe human impact today, but it is a silent crisis – it is a neglected area of research as the climate change debate has been heavily focused on physical effects in the long-term. This human impact report: climate change, therefore, breaks new ground. It focuses on human impact rather than physical consequences. It looks at the increasingly negative consequences that people around the world face as a result of a changing climate. Rather than focusing on environmental events in 50-100 years, the report takes a unique social angle. It seeks to highlight the magnitude of the crisis at hand in the hope to steer the debate towards urgent action to overcome this challenge and reduce the suffering it causes.

The human impact of climate change is happening right now – it requires urgent attention. Events like weather-related disasters, desertification and rising sea levels, exacerbated by climate change, affect individuals and communities around the world. They bring hunger, disease, poverty, and lost livelihoods – reducing economic growth and posing a threat to social and, even, political stability. Many people are not resilient to extreme weather patterns and climate variability. They are unable to protect their families, livelihoods and food supply from negative impacts of seasonal rainfall leading to floods or water scarcity during extended droughts. Climate change is multiplying these risks.

Today, we are at a critical juncture – just months prior to the Copenhagen summit where negotiations for a post-2012 climate agreement must be finalized. Negotiators cannot afford to ignore the current impact of climate change on human society. The responsibility of nations in Copenhagen is not only to contain a serious future threat, but also to address a major contemporary crisis. The urgency is all the more apparent since experts are constantly correcting their own predictions about climate change, with the result that climate change is now considered to be occurring more rapidly than even the most aggressive models recently suggested. The unsettling anatomy of the human impact of climate change cannot be ignored at the negotiating tables.

Climate Change is a Multiplier of Human Impacts and Risks

Climate change is already seriously affecting hundreds of millions of people today and in the next twenty years those affected will likely more than double – making it the greatest emerging humanitarian challenge of our time. Those seriously affected are in need of immediate assistance either following a weather-related disasteror because livelihoods have been severely compromised by climate change. The number of those severely affected by climate change is more than ten times greater than for instance those injured in traffic accidents each year, and more than the global annual number of new malaria cases. Within the next 20 years, one in ten of the world's present population could be directly and seriously affected. Already today, hundreds of thousands of lives are lost every year due to climate change. This will rise to roughly half a million in 20 years. Over nine in ten deaths are related to gradual environmental degradation due to climate change – principally malnutrition, diarrhoea, malaria, with the remaining deaths being linked to weather-related disasters brought about by climate change. Economic losses due to climate change currently amount to more than one hundred billion US dollars per year, which is more than the individual national GDPs of three quarters of the world's countries. This figure constitutes more than the total of all Official Development Assistance in a given year. Already today, over half a billion people are at extreme risk to the impacts of climate change, and six in ten people are vulnerable to climate change in a physical and socio-economic sense. The majority of the world's population does not have the capacity to cope with the impact of climate change without suffering a potentially irreversible loss of wellbeing or risk of loss of life. The populations most gravely and immediately at risk live in some of the poorest areas that are also highly prone to climate change – in particular, the semi-arid dry land belt countries from the Sahara to the Middle East and Central Asia, as well as sub-Saharan Africa, South Asian waterways and Small Island Developing States.

A Question of Justice

It is a grave global justice concern that those who suffer most from climate change have done the least to cause it. Developing countries bear over nine-tenths of the climate change burden: 98 per cent of the seriously affected and 99 per cent of all deaths from weather-related disasters, along with over 90 per cent of the total economic losses. The 50 Least Developed Countries contribute less than 1 per cent of global carbon emissions.

Climate change exacerbates existing inequalities faced by vulnerable groups particularly women, children and the elderly. The consequences of climate change and poverty are not distributed uniformly within communities. Individual and social factors determine vulnerability and capacity to adapt to the effects of climate change. Women account for two-thirds of the world's poor and comprise about seven in ten agricultural workers. Women and children are disproportionately represented among people displaced by extreme weather events and other climate shocks. The poorest are hardest hit, but the human impact of climate change is a global issue. Developed nations are also seriously affected, and increasingly so. The human impact of recent heat waves, floods, storms and forest fires in rich countries has been alarming. Australia is perhaps the developed nation most vulnerable to the direct impacts of climate change and also to the indirect impact from neighbouring countries that are stressed by climate change.

The Time to Act is Now

Climate change threatens sustainable development and all eight Millennium Development Goals. The international community agreed at the beginning of the new millennium to eradicate extreme hunger and poverty by 2015. Yet, today, climate change is already responsible for forcing some fifty million additional people to go hungry and driving over ten million additional people into extreme poverty. Between one-fifth and one-third of Official Development Assistance is in climate sensitive sectors and thereby highly exposed to climate risks. To avert the worst outcomes of climate change, adaptation efforts need to be scaled up by a factor of more than 100 in developing countries. The only way to reduce the present human impact is through adaptation. But funding for adaptation in developing countries is not even one per cent of what is needed. The multilateral funds that have been pledged for climate change adaptation funding currently amount to under half a billion US dollars. Despite the lack of funding, some cases of successful adaptation do provide a glimmer of hope.

Bangladesh is one such example. Cyclone Sidr, which struck Bangladesh in 2007, demonstrates how well adaptation and prevention efforts can pay off. Disaster preparation measures, such as early warning systems and storm-proof houses, minimized damage and destruction. Cyclone Sidr's still considerable death toll of 3,400, and economic damages of $1.6 billion, nevertheless compare favourably to the similar scale cyclone Nargis, which hit Myanmar in 2008, resulting in close to? 150,000 deaths and economic losses of around $4 billion.Solutions do also exist for reducing greenhouse gas emissions, some even with multiple benefits. For instance, black carbon from soot, released by staple energy sources in poor communities, is likely causing as much as 18 per cent of warming. The provision of affordable alternative cooking stoves to the poor can, therefore, have both positive health results, since smoke is eliminated, and an immediate impact on reducing emissions, since soot only remains in the atmosphere for a few weeks. Integrating strategies between adaptation, mitigation, development and disaster risk reduction can and must be mutually reinforcing. Climate change adaptation, mitigation, humanitarian assistance and development aid underpin each other, but is supported by different sets of institutions, knowledge centres, policy frameworks and funding mechanisms. These policies are essential to combat the human impact of climate change, but their links to one another have received inadequate attention.

A key conclusion of this report is that the global society must work together if humanity is to overcome this shared challenge: nations have to realize their common interest at Copenhagen, acting decisively with one voice; humanitarian and development actors of all kinds have to pool resources, expertise and efforts in order to deal with the rapidly expanding challenges brought about by climate change; and in general, people, businesses, and communities everywhere should become engaged and promote steps to tackle climate change and end the suffering it causes.

Conclusions

It can be said that climate change is the antithesis of sustainable development. Sustainable development aims at increasing economic prosperity, safeguarding the environment and improving social equity. As it stands, climate change will impact heavily on the economy – the per year equivalent of the GDP of three quarters of the world's countries already today, according to this report – and is causing millions of people to enter poverty once more. It will cause massive degradation to the environment and human habitat worldwide, including glacial and ice-cap melting, desertification, coastal flooding and soil salination, in addition to much, much more.

Above all, climate change affects the world's poorest first and foremost. 99 per cent of all casualties occur in developing countries. The billions already living on the edge of survival have nowhere to turn when climate change levels its additional burden. Social impacts are diffuse and diverse, people living on two meals a day may be forced to get by on one, basic nutrition or health care are neglected, children may be taken out of school, or women forced into prostitution. In sum, social inequities will become entrenched globally. So climate change undermines the key principles of sustainable development, constituting a serious threat to socio-economic progress worldwide

References

Agrawala, S., *et al.*, 2003. Development and climate change in Bangladesh: Focus on coast flooding and sundarbans." OECD. http://www.oecd.org/dataoecd/46/55/21055658.pdf.

Alcamo, J., Florke, M. and Marker, M., 2007. Future longterm changes in global water resources driven by socio-economic and climatic changes. *Hydrological Sciences Journal*, 52(2).

Anthoff, 2006. Global and regional exposure to large rises in sea-level: A sensitivity analysis. *Check Tyndall Center for Climate Research.*

Black, R., 2008. Demographics and climate change: Future trends and their policy implications for migration. In: Working Paper

T–27, Issued by the Development Research Centre on Migration, Globalisation and Poverty.

Boana, C., 2008. Forced migration online research guide on climate change displacement. *Forced Migration Online.*

Dasgupta, 2007. *The Impact of Sea Level Rise on Developing Countries.* World Bank Publication.

Dercon, S., 1996. Risk, crop choice, and savings: Evidence from Tanzania. *Economic Development and Cultural Change*, April, 44: 485.

Fishcher, G., Shah, M. and Velthuizen, 2002. Climate change and agricultural vulnerability: A special report prepared as a contribution to the *World Summit on Sustainable Development*, International Institute for Applied Systems Analysis, Austria.

Friel, S. *et al.*, 2008. Global health equity and climate stabilization: A common agenda. *Lancet*, 372: 1677–1683

Hope, C. *et al.*, 2007. Some economics of 'Dangerous' climate change: Reflections on the stern review. *Global Environmental Change*, 17: 311–325.

Hope, C., Stern, N. *et al.*, 2008. Reflections on the stern review (1): The case for strong action to reduce the risks of climate change. *World Economics*, 8(1).

IRIN, 2009. Lebanon: Climate change and politics threaten water wars in Bekaa. *Integrated Regional Information Networks*, April 1. http://www.irinnews.org/Report. aspx?Report Id= 82682

Red Cross, 2009. *Case Study Africa.* Red Cross, Red Cross/Red Crescent Climate Guide. http://www. climatecentre.org/downloads/File/RCRC_ClimateG_Africa.pdf.

Renner, M., 2008. *Environment a Growing Driver in Displacement of People.* World Watch Institute, September. http://www.worldwatch.org/node/5888.

Schmidhuber, J. and Tubiello, F.N., 2007. Global food security under climate change. In: *Proceedings of the National Academy of Sciences of the USA*, 104(50): 19703–19708.

Schmidt, S., Kemfert, C. and Hoppe, P., 2009. The impact of socio-economics and climate change on tropical cyclone losses in the

USA. Reg Environ Change. DOI 10.1007/s10113–008–0082–4, January 29.

Smith, D. and Vivekananda, J., 2007. A climate of conflict. International Alert. http://www.reliefweb.int/rw/lib.nsf/db900sid/EMAE–79ST3Q/$file/International per cent 20

Swiss Re., 2009. Natural catastrophes and man-made disasters in 2008: North America and Asia suffer heavy losses." Swiss Re, Sigma Number 2/2009.

Tin, T. Dr., 2007. Climate change: Faster, sooners, stronger. In: *An Overview of the Climate Science Published Since the Fourth IPCC Assessment Report*. World Wildlife Foundation (WWF).

UN, 2007. *Disaster Risk Reduction: Global Review 2007*. United Nations, Geneva, Switzerland.

UN, 2008. *The Millennium Development Goals Report 2008*. United Nations Secretariat, New York.

UNEP, 2006. *The Environment and Security Initiative: Annual Progress Report*. UNEP, UNDP, NATO, OSCE.

UNEP, 2009. *From Conflict to Peacebuilding: United Nations Environmental Programme*. http://postconflict.unep.ch/publications/pcdmb_policy_01.pdf.

UNICEF, 2007. *Climate Change and Children*. UNICEF

WHO, 2008. *Protecting Health From Climate Change*. The World Health Organization–World Health Day 2008. http://www.who.int/world–health–day/toolkit/report_web.pd

Zhang, Y. *et al.*, 2007. Climate change and disability adjusted life years. *Journal of Environmental Health.*

Chapter 10

Climate Change Mitigation: What Do We Do

Rakesh Parmar[1]*, Rakesh Sharma*[2] *and P.B. Reddy*[3]

[1]*Govt. P.G. College, Jaora, (Ratlam)*

[2]*Director, Vision Public School, Nagda*

[3]*Department of Zoology, Nagda, Ujjain*

ABSTRACT

Sustainability is a journey rather than a destination. We will probably never ge tthere–there will always be something we can do better. That hasn't stopped us looking for tangible measures of success. The real measure of our success will be when graduates are empowered to do the right thing in the workplace."

Climate change and its impact on our environment, our economies and our security, is the defining issue of our era. But every day of inaction makes its consequences more irreversible, so we need to act now. To achieve the emission reductions necessary to forestall a continuing cycle of global warming, a broad range of policy instruments will need to be deployed; and action by all countries will be required. The pricing of GHG emissions, directly via carbon taxes or "more probably" through cap-and-trade systems, is an essential element of any comprehensive strategy. Such pricing leads to cost-effective emissions abatement and creates incentives to innovate and deploy new technologies. Cost-effectiveness is essential for creating the necessary consensus for effective climate change policies. In this document, the OECD expands its analysis in two important domains: first, it focuses on the role of technological innovation in bringing down the costs of climate change mitigation over time. It argues that a concerted research and development effort can indeed be expected to yield important benefits, but not by itself. The pricing of GHG emissions is critical for ensuring that new technologies, once developed, are rapidly deployed where they are most needed.

The second new element is an analysis of carbon leakage. that is, the concern that GHG-intensive industries located in countries that take action will lose competitiveness vis-à-vis competitors in countries that do not. The OECD assessment is that while such effects are of concern, they diminish quite rapidly as the set of countries participating in climate change mitigation efforts grow. Excluding energy-intensive industries from GHG emission abatement rules is not economically efficient because it substantially increases abatement costs for the economy as a whole. In the absence of wide country participation in mitigation action, sector-specific arrangements for some of these industries may be an appropriate instrument for reducing carbon-leakage concerns. But it is clearly second best. In the meantime, while we work to restore confidence in the financial markets and to reactivate the world economy, we must integrate in such efforts the need to stop and reverse climate change and promote a "green recovery". Indeed, the current economic crisis can be used to make progress in a new direction, to speed up our efforts to create a new "low-carbon growth" era.

Introduction

Climate change is a fact of life. We need to act urgently if we are to avoid an irreversible build-up of greenhouse gases (GHGs) and global warming at a potentially huge cost to the economy and society worldwide. IPCC analysis suggests that if we act now, we have 10 to 15 years' "breathing space" during which action is possible at a relatively modest cost. But every year of delay reduces this breathing space, while requiring ever more stringent measures to make a difference. Current financial turmoil is not a reason to delay. Indeed, its macro-economic consequences will be resolved in a relatively short time, after which growth will resume, while the consequences of inaction on global warming will continue to grow more and more costly over time.

This review presents some of the major economic issues that will need to be considered in formulating a strategy to address climate change. It is grounded mainly on the report of "The Economics of Climate Change Mitigation" (OECD, 2008) and aims to present some of the key conclusions.

The focus of this review is on what is needed for a cost-effective approach to reducing GHG emissions. Some climate change is unavoidable, due to past and present emissions. Measures will be needed to adapt to this situation: such adaptation policies are explored in depth.

Is a Cost-Effective Climate Change Strategy Possible?

The build-up of greenhouse gases (GHGs) in the atmosphere, much of it driven by human activity, is already affecting the global climate. Under current projections, concentrations of GHGs will continue to increase into the indefinite future, entailing a process of continued global warming. Estimates of the costs of inaction on climate change vary widely, but there is no doubt that beyond a certain level of global warming these costs will be large, particularly in many developing countries as sea-levels and storm surges rise, heat waves become more frequent and intense, and agricultural yields in rural areas decline. Even more disturbingly, each degree of global warming increases the risk of more destructive climate events, causing large and possibly irreversible damage worldwide.

If we look at the costs " and especially the risks " of inaction, ambitious action to reduce GHG emissions makes economic sense. It is vital that such action stabilises GHG concentrations, to slow down and limit global warming.

The IPCC analysis on which this review is based looks at a number of mitigation scenarios. They consider different target levels for long-term concentrations of GHGs, different time frames for achieving them and different ways of compiling the numbers. Any scenario designed to stabilise GHG concentrations at a level that keeps the risks of more destructive climate events moderate is an ambitious one. Most of the long-term targets being discussed would require, for example, eventually reducing GHG emissions to one-third or one-quarter of 2005 levels over the long term. Such radical change transformation of the economy that will not be either easy or cheap, and there is thus an enormous premium on developing a cost-effective set of policy instruments to achieve the required abatement.

This review first offers an overview of where we are, where we are headed and where we need to go as regards climate change. It then focuses on three core issues for reducing global GHG emissions:

- To be cost-effective, cuts in emissions need to be made where they are cheapest. More specifically, the costs of any additional emission cut need to be equal for all sources of emissions, so that it would not be possible to lower overall costs by redistributing cuts. Putting a price on GHG emissions is the most obvious way to achieve this. Equally important, such pricing establishes incentives to undertake research and development (R&D) and innovate more generally in energy-saving and climate-friendly technologies. Such innovation is the best hope for containing the costs of curbing GHGs in the long run. However, in practice a broad mix of policy instruments will have to be deployed, because price measures cannot solve all problems effectively and the coverage of pricing schemes may be less than complete for some time.
- One concern that arises when some countries take ambitious action to abate GHG emissions but others do not is that energy-intensive sectors in abating countries will see the costs as too high for them, due to loss of competitiveness vis-à-vis countries that do not take action. The amount of "leakage" of emissions to countries which are not taking part in the effort appears to be relatively small if the pool of abating countries is reasonably large. However, short-term concerns in these countries about the impact of job-losses in energy-intensive sectors on overall employment could hamper progress in implementing climate policies requires and this booklet examines measures to address these concerns.
- The costs of both climate change and abatement action are unevenly distributed across regions and sectors. For this

reason, incentives to participate in an abatement framework are also heterogeneous. A range of approaches and instruments may be helpful in providing support for action and generating buy-in, thereby achieving the wide country coverage required for cost-effectiveness.

Where we Now and Where are we Headed?

World GHG emissions have roughly doubled since the early 1970s and on current policies could rise by over 70 per cent during 2008-2050. Historically, energy related GHG emissions were predominantly from the richer developed countries, so that the rise in GHG concentration from the industrial revolution to today is largely accounted for by economic activity in these countries. Today, however, two-thirds of the flow of new emissions into the atmosphere is accounted for by developing countries, outside and without new policies this share is set to rise further to 2050.

Greenhouse gases are emitted by many economic activities. Quantitatively, the largest share is accounted for by power generation (electricity production and transformation were responsible for 26 per cent of global emissions in 2004), followed by industry generally (about 19 per cent) and transportation (13 per cent). It is important to note that deforestation and forest degradation (about 17 per cent) are estimated to account for more emissions globally than the entire transport sector. Figure 10.3 illustrates the likely build-up of GHG concentrations to 2100 based on a "business as usual" scenario and without any climate policies beyond those currently in place. It also shows the ultimate extent of global warming associated with these GHG concentrations. The OECD baseline scenario assumes world economic growth averaging just over 3.5 per cent in purchasing-power-parity terms up to 2050, with a gradual catch-up in living standards of developing countries to those of the developed ones. In terms of emissions and resulting concentrations, the baseline is quite close to the average of other recent studies; some are more optimistic, but others less so.

There are clearly grounds for concern. The mid-range estimate for global warming to the end of the century, close to 4° Celsius compared with pre-industrial levels, brings temperature to a level where destructive climate events are not improbable; and the temperature would keep rising into the next century even if

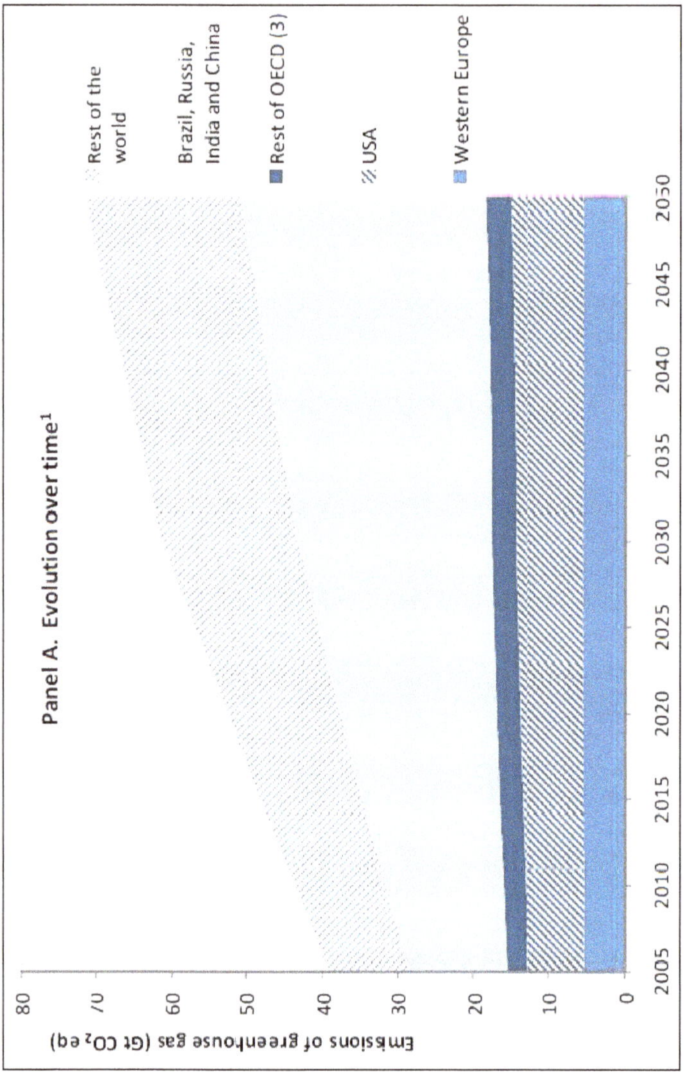

Figure 10.1: Developing Countries would Account for Most of the Projected Increase in World Greenhouse Gas Emissions Over the Coming Decades

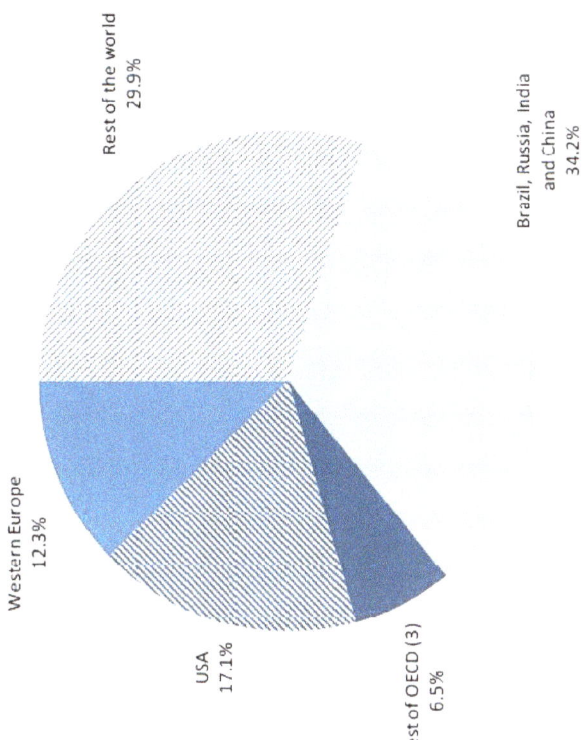

1. Excluding emissions from Land Use, Land-Use Change and Forestry.
2. Including emissions from Land Use, Land-Use Change and Forestry.
3. Rest of OECD does not include Korea, Mexico and Turkey, which are aggregated in Rest of the World.

Figure 10.2

(Projected trends in greenhouse gas concentration and associated temperature increases in the absence of new climate change policies)

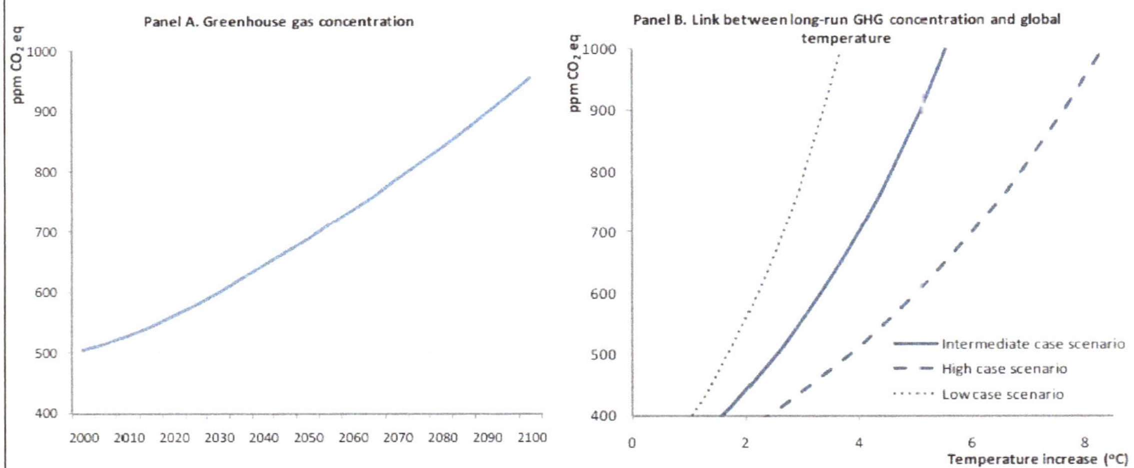

Note: The scenarios embed different values for the climate sensitivity parameter. This parameter measures the impact on temperature of a doubling of concentration and determines the link between long-run GHG concentration and global temperature at the steady state. Because of the inertia of the system, steady-state temperatures may be reached several decades after concentration stabilisation. The climate sensitivity parameter equals 4.5 in the "high case" scenario, 3 in the "intermediate case" scenario, and 2 in the "low case" scenario.

Figure 10.3: Greenhouse Gas Concentration would Rise Sharply without New Policy Action

concentration were to stabilise at that point. It has to be noted that in many respects the OECD baseline scenario is not a particularly pessimistic one. Energy intensity–the amount of energy used to produce a given level of gross domestic product (GDP)–is assumed to fall continuously, partly because as living standards across the world converge, GDP shifts steadily towards less energy-intensive products. At the same time, ongoing efficiency gains are expected in the use of energy, partly some might argue that the problem just cannot be this bad, and at least two lines of argument deserve brief comment here. One is that the rise in oil and other energy prices in recent years should, if sustained, trigger major changes in behaviour that would in turn curb emissions. The second is that the baseline scenario underestimates the capacity for technological innovation to find solutions to problems.

There is of course an important element of truth in the energy price argument, although volatile oil prices do not provide clear signals to motivate long-term changes in behaviour. The price of oil is projected to keep rising as conventional oil becomes increasingly scarce relative to demand because of the depletion of oil reserves. But while oil is scarce, other hydrocarbons "notably non-conventional oils such as the Canadian tar sands and coal" are much less so. So the IPCC baseline shows the price of coal, in particular, as levelling off even as more coal is being used. The problem is that non-conventional oil and coal are actually more carbon-intensive than petroleum, so that the estimated CO_2 intensity per unit of energy consumed rises by some 35 per cent between now and 2050. In other words, any possible reduction in emissions due to less use of motor transport due to high oil prices is mostly offset by a shift to more CO_2-intensive coal in power generation.

When it comes to the argument that innovation will save the day, it is certainly true that innovation policies will play a central role in making a mitigation strategy affordable. But Research and Development (R&D) efforts, however intensive, cannot by themselves accomplish enough if the incentives to deploy new technologies are weak. Figure 10.4 seeks to illustrate this point by presenting an alternative scenario that includes an aggressive world-wide policy of investing in abatement-related R&D, raising such spending four-fold from current levels as a share of global GDP back to the peak levels observed in the early 1980s, but without applying any other policies to reduce emissions.

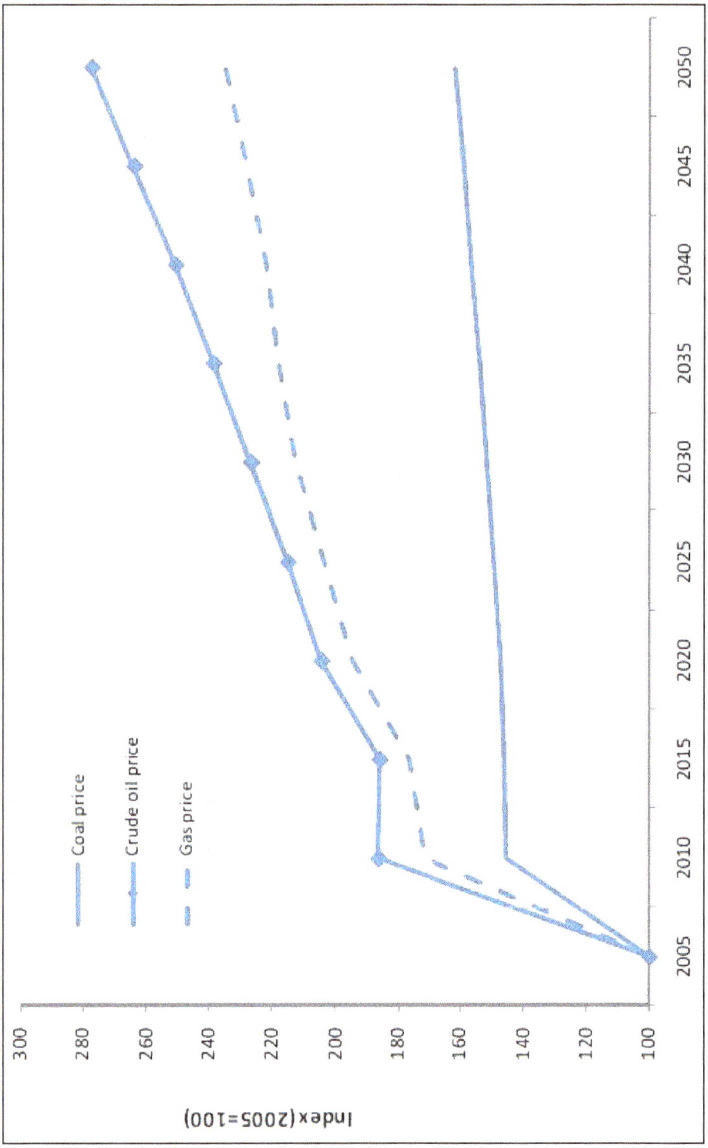

Figure 10.4: The Conventional Oil Price is Set to Rise but Coal would Lag

While such an R&D-led approach clearly yields some results, OECD estimates suggest that these are not nearly enough to stabilise GHG concentration levels by the end of the century. In fact, even a 30-fold increase in worldwide public R&D spending on low-carbon technologies, raising it to about 1 per cent of global GDP annually, would not be enough to achieve this target. The essential problem is that, even if innovative and effective technologies emerge, they will not be intensively used until their costs come down close to those of existing competing technologies, unless incentives are put in place for people to use them. Such cost falls typically take a very long time. Carbon capture and storage (CCS) technology provides a telling and extreme example of this challenge. This technology is still new and expensive. Intensive R&D efforts would in all likelihood bring costs down steadily. But this technology would not be widely used in power generation unless there were significant incentives to cut carbon emissions, since otherwise it reduces generating efficiency and hence raises costs without providing any return.

What can be Done?

Faced with the consequences and costs of inaction, governments have reached a consensus internationally that global emissions need to be cut significantly. Countries are working towards an international framework for action, with the aim of reaching agreement at the UN Climate Change Conference to be held in Copenhagen in late 2009.

The IPCC has simulated a number of possible scenarios for ambitious reductions in GHG emissions along with their economic and environmental impacts compared with those of the baseline scenario with no new policy action.

For illustrative purposes the economic issues involved in designing a comprehensive approach to reducing emissions are explored here primarily by examining one mitigation reference scenario. This scenario eventually stabilizes GHG concentrations at a level equal to 550 parts per million (ppm) of CO_2 (or about 450 ppm CO_2 only), while allowing a moderate overshooting of this level over an interim period. It has to be emphasised that this scenario has no specific normative significance. Indeed, many countries have taken the view that a more ambitious objective would be appropriate as would be required, for instance, to cap the extent of global warming at 2°Celsius. This review focuses on one scenario and uses this to examine quantitatively how different policy assumptions might affect overall results.

Major changes in behaviour and production methods will be needed to achieve GHG mitigation at the lowest possible cost. Mitigation is achieved by reducing both the energy intensity of GDP and the carbon intensity of energy used. As a side effect of these changes, GDP growth will also be affected. Under the mitigation reference scenario examined here, it is estimated that the average growth of the world economy over the period 2008-2050 would be some 0.13 percentage points lower than in the absence of climate change policies. The impact on GDP growth is small in the early years, but increases significantly after 2025. By 2050, the level of world GDP is estimated to be some 4.8 per cent below what it would be in the absence of climate change policies. The reason for this GDP loss is that substantial human and capital resources will have to be shifted to working on GHG mitigation, thus reducing the resources available for producing other goods and services. While abatement obviously generates other benefits in terms of avoided climate change damage, such benefits are not always directly captured in conventional GDP, and are not reflected here. This review suggests that a wide range of instruments will be needed " an effective global strategy should ideally meet a number of core requirements. These can be summarised as follows:

- ✰ Comprehensiveness both in terms of addressing all GHG gases and in engaging the largest possible number of countries and sectors, including especially the largest emitters.
- ✰ Some degree of de-coupling between where abatement takes place and who pays for it, to ensure emission reductions take place where they are cheapest, even if there are limited resources to pay for the reductions locally. No doubt this represents an important challenge for international diplomacy; but it is clear that, at least for the world's poorest countries, support through such a de-coupling will be required.
- ✰ The strategy has to be both credible and flexible. Credibility is essential because no strategy, no matter how well designed, will elicit the required behavioural responses if

people lack confidence that it will be carried through. Flexibility is also needed, because uncertainty margins about both the economic and environmental results of any strategy remain large, so that mid-term corrections to any strategy will almost certainly be required. Building-in from the start such a capacity to make periodic corrections would reduce risks of deadlock that might arise if all changes had to be negotiated along the way. There is no doubt some element of trade-off between credibility " which requires clear targets–and flexibility, which requires the capacity to adjust them. Combining strongly-anchored long-term objectives with rolling shorter-term commitment periods may be one way of balancing this trade-off.

☆ Finally, the strategy needs to be put into place urgently. The illustrative mitigation reference scenario considered in this booklet suggests that even under an ambitious abatement strategy, the required implicit price of carbon emissions is relatively modest over the next 10 to 15 years.

This creates a "breathing space", but this breathing space will be used productively only if the strategy provides an immediate and clear signal for the longer term. Such a breathing space is extremely important for two reasons. First, it provides time for long-lived investment decisions to be made now so that the abatement capacity is in place when it is needed. Second, it avoids a "price-shock" that would in fact be required if the strategy is delayed, and which could well make it much more difficult for governments vis-à-vis their electorates to buy into the strategy. Every year of delay reduces this breathing space, and at the same time raises the stringency of the measures required to meet the objective. In this double sense, delay is extremely costly.

References

Barlow, Maude, 1999. *Blue Gold: The Global Water Crisis and the Commodification of the World's Water Supply.* International Forum on Globalization, June, Sausalito, Calif.

Boulding, Kenneth E., 1993. The economics of the coming spaceship earth. In: *Valuing the Earth: Economics, Ecology, Ethics,* (Eds.) Herman E. Daly and Kenneth N. Townsend. MIT Press, Cambridge, MA, p. 297–309.

Cosgrove, William J. and Rijsberman, Frank R. for World Water Council, 2000. *World Water Vision.* Earthscan and World Water Council, London.

Costanza, Robert, Low, Bobbi S., Ostrom, Elinor and Wilson, James (Eds.), 2001. *Institutions, Ecosystems and Sustainability.* Lewis Publishers, Boca Raton, FL.

Daly, Herman E., 1993. The steady: State economy. In: *Valuing the Earth: Economics, Ecology, Ethics,* (Eds.) Herman E. Daly and Kenneth N. Townsend. MIT Press, Cambridge, MA, p. 325–363.

Dryzek, John S., 1987. *Rational Ecology: Environment and Political Economy.* Basil Blackwell, New York.

Georgescu-Roegen, Nicholas, 1993. Selections from "energy and economic myths." In: *Valuing the Earth: Economics, Ecology, Ethics,* (Eds.) Herman E. Daly and Kenneth N. Townsend. MIT Press, Cambridge, MA, p. 89–112.

Gleick, Peter H., 1993. Water and conflict: Freshwater resources and international security. In: *International Security* 18: 79–112. 1998 and 2000. *The World's Water: The Biennial Report on Freshwater Resources.* Island Press, Washington, DC.

Herring, Horace, 2002. Is energy efficiency environmentally friendly? *Energy and Environment,* 11: 313–325.

International Joint Commission, 1992. *Sixth Biennial Report on Great Lakes Water Quality.* Windsor, Ontario.

Jevons, Stanley W., 1865. *The Coal Question: An Inquiry Concerning the Progress of the Nation, and the Probable Exhaustion of Our Coal Mines.* Macmillan, London.

Postel, Sandra, 1999. *Pillar of Sand: Can the Irrigation Miracle Last?* Norton, New York.

Postel, Sandra L., and Wolf, Aaron T., 2001. Dehydrating Conflict. *Foreign Policy,* September.

Princen, Thomas. 1994. Toward a Theory of the International Political Economy of Sustainability. *International Studies Notes* 19: 40–43

Princen, Thomas, Maniates, Michael and Conca, Ken, 2002a. Conclusion: To confront consumption. In: *Confronting Consumption,* (Eds.) Thomas Princen, Michael Maniates and Ken Conca. MIT Press, Cambridge, Mass.

Princen, Thomas, Maniates, Michael and Conca, Ken, 2002a (eds). 2002b. *Confronting Consumption.* MIT Press, Cambridge, Mass, p. 317–328.

Rudin, Andrew, 2002. DSM: Ten Years Later. *Engineered Systems* May, p. 80–82.

Chapter 11

Climate Change and its Impact

Dilip Gupta
Research Scholar,
R.B.S. College, Agra

ABSTRACT

Everyone know about the mystery of climate change. Climate change is not a new matter. Increase earth's average temperature is called "Global Warming". Climate change is used to describe a change in climate. In the context Climate is taken to mean the average weather. Climate can change over period of time ranging from months to thousand or millions of year. So this is a big agenda in present time. The causes of climate change are human activities, effect 90 per cent on it. The fossil fuels, green house gases (by the uses of Refrigerator, Air Conditioners, Aerosols) are the main cause to change the climate. In present time industrial area growing very fast due to cause global warming.

The effect on Biodiversity, Agriculture, Sea level changes, Increasing floods and weather disasters, Increasing the acidity of ocean water, Increasing probability in drought and fires are great symbol of climate change. By the causes of climate change increase more power of tropical cyclones. Every year waste millions dollars by the causes of tropical cyclones. A large quantity of population lives in developing countries which is depend on fossil fuels to their needs. The productivity of Agriculture is decreasing so that is the straight effect of developing countries population.

The ecosystem is a weak step of environment. If any change in climate, fast effect on ecosystem. Many species of Marine doing struggle to save their life. The same effect is as on terrestrial ecosystem also. Scientist all over the world are making prediction about the ill effect of climate change. So we wants that all human of the world needs to join this agenda to save our planet.

Introduction

Increase earth's average temperature is called Global Warming. "Climate Change" is used to describe a change in the climate. In the context "Climate" is taken to mean the average weather. Climate can change over period of time ranging from months to thousand or millions of years. The classical time period is 30 years, as defined by the world metrological organization. The climate change referred to may be due to natural causes, *e.g.* Changes in the sun's output or due to human activities *e.g.* changing the composition of the atmosphere. Different parts in countries many climate specialists studying the climate record of the earth have the same opinion now that human actions, mainly the discharge of green house gases from smoke stacks, vehicles, and burning forests. By the uses of traditional fuels cause the 90 per cent of green house gases. In developing countries use huge quantity of traditional fuels. In present time industrial areas growing very fast due to cause global warming. The chimneys of factory, we called that "Man made Volcanoes" support Global warming. The major cause of global warming is the emission of green house gases like carbon di-oxide, methane, nitrous oxide etc. into the atmosphere. The major source of carbon di-oxide is the power plants. These power plants emit large amounts of carbon di-oxide produce from burning of fossil fuels for the purpose of electricity generation. About 20 per cent of carbon di-oxide emitted in the atmosphere comes from burning of gasoline in the engines of the vehicles. This is true for most of the developed countries building; both commercial and residential represent a larger source of global warming pollution than cars and trucks. Methane gases which called the marsh gas are more than 20 times as effectual as CO_2 at entrapping heat in the atmosphere. Methane is obtained from resources such as rice paddies bacteria in bogs and fossil fuel manufacture. When fields are flooded, anaerobic, situation build up and the organic matter in the soil decays, releasing methane to the atmosphere. The main sources of nitrous oxide include nylon and nitric acid production, cars with catalytic converters, the use of fertilizers in agriculture and the burning of organic matter. Another cause of global warming is deforestation that is caused by the cutting and burning of forest for the purpose of residence and industrialization our worried on climate change is as:

- ☆ The concentrations of green house gases grow fast with time cause average temperature also grow up.
- ☆ The movement of climate change is so fast than the human life history.
- ☆ The 90 per cent cause of Global Warming due to human activities man natural powers.

Objectives

The paper aims to:

1. Analyse the losses of climate change.
2. Study the impact of climate change on atmosphere and human infrastructure.
3. Appraise the various effects on developing countries.
4. Indentify the path way to make our progress system sustainable.

Databse and Methodology

The data used for the analysis has been drawn from a wide variety of sources life the IPCC (Intergovernmental Panel on Climatic Change) World Meteorological Department, Transport Research wing etc. The methodology incorporates computation of data and its analysis to infer how rapidly we are using fossil fuels and industrialization and the impact of the world climatic regions on the physical and social environment and infrastructure. The study area belong the world climate.

Climate Change is Inspiring Scientists to Fight for Awareness

Scientists all over the world are making predictions about the ill effects of climate change and connecting some of the events that have taken place in the pat few decades as an alarm of global warming. The effects of global warming are increasing the average temperature of the earth. A rise in earth's temperature can in turn root to other alteration in the ecology, including an increasing sea level and modifying the quantity and pattern of rainfall. These modifications may boost the occurrence and concentration of sever climate events, such as floods, famines, heat waves, tornado, and twister. Other consequences may comprise of higher or lower agricultural outputs, glacier melting, and raise the ranges of disease vectors. As an effect of global warming species like golden toad, harlequin frog of Costa Rica has already become extinct. There are number of species that have a threat of disappearing soon as an effect of climate change. As an effect of climate change various new diseases have emerege lately. On increase the earth average temperature, the bacteria can survive better in elevated temperatures and even multiplies faster when the conditions are favourable many types of Mosquitoes due to increase in humidity levels and their frequent growth in warmer atmosphere.

A group of scientists have recently reported on the surprisingly speedy rise in the discharge of carbon and methane release from frozen tundra in Siberia now starting to melt because of human cause increase in earth's temperature.

Various Impacts of Climate Change

Impact on Biodiversity

The major effect of climate change has been noticed in the behavior of the wildlife. The various species lost due to climate change. Major changes are seen in the animals. A behaviour pattern of the animals studied shows that the animals are beginning to shift their population towards north or towards a higher altitudes. The polar bear of Arctic regions will die due to climate change. Marine life is also very sensitive to increase in temperatures. The effect of climate change will definitely be seen on some species in the water. A survey was made in which the marine life reacted significantly to changes in water temperatures. It is expected that many species of fish as (prone, salmon) will die off or become extinct due to the increase in the temperatures of the water. It's true that the most disturbing changes are expected in the coral reefs that are expected to die off as an effect of climate change. Even the penguins that live on the poles are getting affected as a result of climate change. Due to melting of polar ice their life cycle is getting disturbed and which leads to death of many animals. Coral reefs, the house of many types of marine species of fish, which are highly sensitive to small changes in water temperature suffered the worst bleaching or die off response to stress ever recorded in 1998 with some areas seeing bleach rates of 7 per cent. Experts expect these sorts of events to increase in frequency and intensity in the next 50 years as sea temperature. Some species life golden toad, harlequin frog of Costa Rica had already become extinct. By the causes of climate change. The ecosystem of terrestrial and marine change very fast.

Impacts of Agriculture

According the William R. Cline the climate change can affect agriculture in a variety of ways. Beyond a certain range of temperature, warming trends to reduce yields because crops speed through their development producing less grain in the process and higher temperature also interfere with the ability of plants to get and use moisture. Evaporation from the soil accelerates when temperatures rise and plants increase transpiration that is; lose more moisture from their leaves. The combined effect is called "Evapotranspiration" because global warming is life to increase rainfall the net impact of higher temperatures on water availability is a race between higher evapotranspiration and higher precipitation. Typically that race is won by higher evapotranspiration.

It steps are not taken to curb carbon emission agricultural producing could fall dramatically especially in developing countries. Agriculture is a larger share of developing economies than of economies.

Damage will generally be greater in countries located closer to equator, where temperature already tend to be close to crop tolerance levels. The most severely affected countries are in Africa, Latin America and South Asia.

The developing countries are most at risk, so it is strongly in their own interest that they participate actively in international abatement programs. China could be less interested in international efforts to rest rain emission than India, which face major potential losses if there is no change in global emissions policies.

Impacts on Ocean

The oceans serve as a sink for carbon di-oxide, taking up much that would otherwise remain in the atmosphere, but increased levels of CO_2 have led to ocean acidification. Move than 90 per cent of warming that occurred over 1960-2010 is estimated to have gone into the oceans. As carbon di-oxide dissolved in sea water, carbonic acid in formed which has the effect of acidifying the ocean, measured as a change in pH. The effects of ocean acidification on the marine biosphere. Phytoplankton, the base of the marine food chain will destroyed by the effect of global warming may marine fishes and other creature of ocean will destroyed because phytoplankton provide food to them. Many coral reefs will also effect to large. Many types of fishes built their nest on coral reefs. Many countries in world depend their food on sea food, so that, this crisis is so much big. Melting ice sheets and glaciers is a big caution of may countries and islands, which is low high of the sea level on happen that's position create a big migration on continents. The scientists observe the temperature of Antarctica rose by 0.17°C between the 1950's and the 1980's, nearly twice the rate for the world is ocean as whole. Other possible effects include large scale changes in ocean circulation.

Effects on Climatic Change on Human, Industry and Society

Human beings are exposed to climate change through changing weather patterns and indirectly through changes in water, air, and food quality and changes ecosystem, agriculture, industry and settlements and the economy. A study by the WHO 2009 estimated the effect of climate change on human health. Climate change responsible for 3 per cent of diarrhoea 3 per cent of malaria and 3.8 per cent of dengue fever deaths world wide in 2004. These increases were primarily in low income population already experiencing a large burden of disease.

Floods and Weather Disasters

Floods are low probability, high impacts events that can over whelm physical infrastructure and human communities. Floods and tropical cyclones have the greatest impact in South Asia and Latin America. High density population in low lying coastal regions experience a high health burden from weather disaster.

Drought and Fires

The effects of droughts on health include deaths, malnutrition, infections disease and respiratory diseases. In some regions changes in temperature and precipitation are projected to increase the frequency and severity of fire events. Forest and bush fires cause burns, damage from smoke inhabitation and other injuries.

Effect on Water Resources

A number of climate related trends have been observed that effect water resources. These include change in precipitation and surface water sea level rise will extend areas of salinisation of groundwater and estuaries, resulting in a decrease in freshwater availability for humans and ecosystems in coastal areas IPCC fourth assessment report (Africa, Asia, Australia and New Zealand, Europe, Latin America, North America, Polar regions and small islands) showed an overall net negative impact of climate change on water resources and freshwater ecosystems.

Effecton Economy

Climate change impacts economy various types. Its great effects on developing country because developing country economy based on agriculture. Health impact, vegetation impacts are examples of climate change.

Great Effect of Climate Change in Author's View

By the concept of plate tectonics theory earth surface is made of rigid body, called plate. Its average depth is 100 km which include crust and upper mental. The movement of plate cause deep earth quake and much other change on earth. On melting ice a large quantity of water will flow in the sea. The quantity of water pressure on the bottom of ocean, which help moving the tectonic plate so I want to say that the frequency of earthquakes are increase that is the big disaster for developing and developed country.

Conclusion

Population is big disaster for every developing country. The demand and food supply is connected to climate change process because developing country population based on fossil fuels to cook their food and many other needs.

Tree plays main role to reduce pollution and maintain climatic changes. It changes CO_2 into O_2 which provide us life so increase tree revaluation with green revaluation. CO_2, methane, SO_2 are green house gases which are the main cause of climate changes so we wants that to reduce these gases in environment by the uses of Anti Green House Gases Products.

References

Cline, William R., 2007. *Global Warming and Agriculture.*

Lal, D.S., 2008. *Climatology and Oceanography*. Sharda Pustak Bhawan, Allahabad.

Singh, Savindra, 2004. *Environmental Geography*. Vasundhra Prakashan, Gorakhpur.

Singh, Savindra, 2004. *Physical Geography*. Vasundhra Prakashan, Gorakhpur.

www.globalwarmingandglaciers

www.ipccindex.htm2009

Chapter 12

Forest Resources in a Metropolitan City: Delhi

Chitra Sejwar[1]* *and B.S. Rathor*[2]

[1]*Associate Professor, Department of Geography, J.S. (P.G.) College, Sikandrabad, C.C.S. University, Meerut, U.P.*

[2]*Head, Department of Geography, Govt. P.G. College, Morena, M.P.*

ABSTRACT

Delhi with a population of 12.19 million (2001) is the third largest city of India. Delhi forests have acquired immense importance now-a-days due to their ecological role in maintaining environmental stability, their role in combating high pollution level and mitigating climate change in Delhi. Presently, there is 299.58 sq. km of forest and tree cover in N.C.T. of Delhi which is about 20.20 per cent of the Geographical area. Delhi Ridge Forests acts as a "Green Lung" to Delhiites. This study is undertaken keeping in view the role of Delhi forests in mitigation of climate change and combating pollution.

Keywords: *Conservation, Afforestation, Ecological and Environmental stability.*

Introduction

Food and Agricultural Organization of United Nation (FAO) defines forest as land with a tree canopy cover of more than

* Corresponding author: E-mail: drchitrasejwar@yahoo.co.in

10 per cent and area of more than 0.5 hectare. A forest is a biotic community which is predominantly composed of trees, shrubs or any woody vegetation and often with a closed canopy. They are storehouse of biodiversity and provide important environmental services to mankind.

The forest performs the productive, protective and regulative functions for mankind. The productive functions of forest include the production of timber, bamboo's, food and a wide variety of compounds. Such as resins, latex, alkaloids etc. The protective functions include conservation of soil and water, prevention of drought and protection against wind, cold, noise etc. The regulative functions of forests improve atmosphere and temperature conditions, it also enhances the economic environmental value of landscape. Forest combat pollution, sequester carbon and play an important role in mitigating climate change. Therefore in view of above cited facts and with the growing realization of the impacts of forests on climate change and its role in combating pollution the present study is undertaken.

Objectives

The objectives of the instant research paper are as follows:

- ☆ To study spatio-temporal changes of forests and tree cover in Delhi.
- ☆ To study the role of forest in combating pollution in Delhi.
- ☆ To assess the forest conservation and management practices in Delhi.

Study Area

Delhi lies between 28°24'17" and 28°30'0" North latitude and 76°50'24" and 77°20'37" East longitudes. It is surrounded on three sides, *i.e.*, on north, west and south by Haryana and to the east across the river Yamuna by Uttar Pradesh. The major part of Delhi lies on the western side of river Yamuna, while only some villages and urban areas of north-east and eastern districts lies on the eastern side of the river. Its greatest length is about 51.90 km and the greatest breath is 48.48 km. Delhi's altitude range between 213 and 305 metre above mean sea level. It covers 1,483 sq.km area (558.32 sq.km rural area and 924.68 sq. km urban area). According to 2001 census the total population of Delhi was 1,38,50,507 persons. The geographical area of Delhi is like a board alluvial neck, bounded by Thar Desert on the south, by mighty Himalayas on the north, Indo-gangetic plains on the north east, the fertile gangetic plains on the east and by Aravalli hills on the south. The topography of Delhi has two important physical features- Ridge and river Yamuna. The average annual rainfall ranges between 400-600mm.The average annual temperature ranges between 22° to 45°celsius.

Data Sources and Methodology

The secondary is data taken from 'India–State of Forests Report' from the year 1993 to 2009. It deals with spatio-temporal aspects of forests and tree cover in Delhi. The sample unit is district. The data has been analyzed and statistical data has been shown with the help of tables, maps and bar diagrams.

Forest Resources of Delhi

The natural forests of Delhi are mainly of tropical dry deciduous type and are a thorny Shrub. In the classification of forest types developed by Champion *et al.* (1968), the forests of Delhi fall in the category of tropical thorn forests, and more specifically as semi-arid open shrub of dakh forests. The overall forest cover of Delhi is divisible into two general types- Natural and Plantation. The natural forests generally include Kikar and thorny bushes. Forests under afforestation schemes include temperate, tropical and deciduous trees which are grouped under mixed forests.

The forests of Delhi have been broadly grouped into two categories *i.e.*, reserved and protected forests which are briefly described as below.

Reserve Forest

An area so constituted under the provisions of the Indian Forest Act or other State Forest Acts, having full degree of protection. In Reserved forests all activities are prohibited unless permitted. Main Reserve Forest in Delhi.

Ridge Forest

A spur of the Aravalli hills of Rajasthan, spreads across the territory of Delhi in the South- West, North-East direction is called Delhi Ridge. The Ridge and its neighbouring hilly tracts represent the natural flora. The major natural forests in Delhi are generally restricted to the Ridge. The natural flora is a tropical, thorny and secondary forest. The greater of the reserved forest of this territory is located in the New Delhi Ridge, opposite to the Rashtrapati Bhawan at the back of Laxmi Narain Temple. The Vegetation of the Ridge can be grouped into two broad categories which are:

1. Permanent Vegetation
2. Ephemeral Vegetation

The Permanent Vegetation of the Ridge area abounds in thorny tree species and a few shrub species like *Kikar, Katha, Ronj, Phulahi.* The chief shrubs are *Heens, Ber, Lasoora* and *Bansa* etc.

The ephemeral vegetation of the Ridge area comprises the annuals, which come up mainly during the brief monsoon season when moisture is abundant both in the soil and atmosphere. Some of the common herbs of the ephemeral vegetation include *lapa, missi, till, punarnava, gokhru.* The thorny shrubs of the permanent vegetation provide support to the growth of many climbers and twiners such as *Amarbel, Rembel etc.*

The ridge achieve the maximum elevation of 316 m near Bhatti, mines relatively this point is about 10m higher than Yamuna Railway bridge. The ridge forms about 6 per cent of the total land area of Delhi.

The ridge consists of four relatively small pockets of forests. The total area now actually available in the ridge is approximate 7777 hectare which is not a continuous stretch but divided into four zones namely.

1. *Northern Ridge or the old Delhi Ridge*: About 87 hectare in area, it is the smallest section bounded by Delhi University in the north. Kamla Nagar in the south, Rajpur Road in the east and Malka Ganj in the west. It is also known as Kamla Nehru Ridge.
2. *The Central or the New Delhi Ridge*: More popularly known as Dhaula-Kuan Ridge is 864 hectare is bordered in the north, south, east and west by Link road, Dhaula-Kuan, Mandir Marg and Narayan Industrial area respectively.
3. *The South Central Ridge (Vasant Kunj)*: It is about 626 hectare in area and is flanked by Mehrauli Institutional area,

Mahipalpur, Sri Aurobindo Marg and Jawhar Lal Nehru University road in north, south, east and west respectively.

4. *The Southern Ridge*: About 6200 hectare in area, where a substantial tree cover is found in the Sanjay Nana or Kishan Garh forests and the Jahanpanah City Forests. The remainder of the Southern ridge is mostly the Asola-Bhatti Wild Life Sanctuary which is protected but is largely degraded owing to mining and grazing activities in the past.

In May, 1994, the Lt. Governor of Delhi notified the entire ridge, on the basis of present day boundaries as Reserve forest under Section 4 of Indian Forest Act- 1927, in accordance with recommendations of the Lavraj Kumar Committee.

Protected Forest

An area notified under the provisions of the Indian Forest Act or other State Forest Acts, having limited degree of protection. In protected forest all activities are permitted unless prohibited. Main protected forests in Delhi are as:

Sanjay Van

It lies next to CRPF Camp between Lalkot and Qutub Institutional Area. Sanjay Van still represents the original ridge forests having native species like *Kadam and Siras* trees. It has a good population of native fauna and is recorded that *Neelgai* is more common here than any other part of the ridge. The diversity of bird species is also extensive.

Asola-Bhatti Wildlife Sanctuary

On Oct, 9th 1986, the Governor of Delhi declared 1880 hectare. of land in the Southern ridge as Asola Wild Sanctuary. Further to this, in April, 1991, the adjoining 840 hectare. Of Bhatti Mines were brought under Asola Wild Life Sanctuary. It is located near Chhatarpur, Tughlaqbad Bhatti mines in Mehrauli Block and forms a part of Southern ridge with overall area of 6200 hectare. The land is generally degraded and devoid of trees other than Kikar. These Asola- Bhatti Wild Life Sanctuary supports a good population of Neelgai, occasional fox and jackal. About 30 check dams, harvesting rainwater have changed the face of this place over the last few years. Thus indigenous species like anogysis, pendula, dhak, jungle jalebi are reappearing in the Asola forest.

Forest and Tree Cover in Delhi

According to *India-State of Forest Report–2009* of the Forest Survey of India, Dehradun, Total forest and tree cover is 20.20 per cent of the state's Geographical area of 1483 sq. km as shown in Table 12.1. Delhi has more forest cover area than the cities of Bangalore, Mumbai, Kolkata and Chennai. District wise Forest cover of Delhi is shown in Table 12.1 also the change in forest and tree cover in Delhi is shown in Table 12.2. From Table 12.1 it is seen that there is maximum forest in South Delhi and minimum in East Delhi. From Table 12.2 it is seen that there is increse in forest cover from 1993 to 2007 but there slight decline in the year 2009 due to various infrastruture projects.

Table 12.1: District-wise Forest Cover of Delhi (Area in Sq. km)

Sl.No.	*District*	*Geographic Area*	*Forest Cover*		
			2001	*2003*	*2007*
1.	Central Delhi	24.68	2.27	5.2	5.05
2.	East Delhi	63.76	1.57	2.94	2.99
3.	New Delhi	34.90	9.73	14.54	16.31
4.	North Delhi	59.16	3.18	4.7	4.81
5.	North East Delhi	60.29	1.85	2.7	3.59
6.	North West Delhi	440.31	8.37	15.47	16.41
7.	South Delhi	249.85	52.51	40.61	78.64
8.	South West Delhi	420.54	30.09	78.65	42.45
9.	West Delhi	129.52	1.76	5.06	6.33
	Total	1483.00	111.33	139.87	176.58

Source: State of Forest Report 2005. p-52, 2009 p-76.

It is seen from the Figure 12.2 that South Delhi, South-West Delhi and New Delhi have more forest cover than the rest of Delhi and East Delhi has minimum forest in Delhi.

Role of Forest in Combating Pollution

Vegetation cleans the air and along with water resources, acts as a moderator of climate. One hectare of woodland (about 1000 trees) absorbs 3.7 tonnes of carbon di-oxide from the atmosphere and gives out 2.5 tonnes of life-sustaining oxygen. A full grown

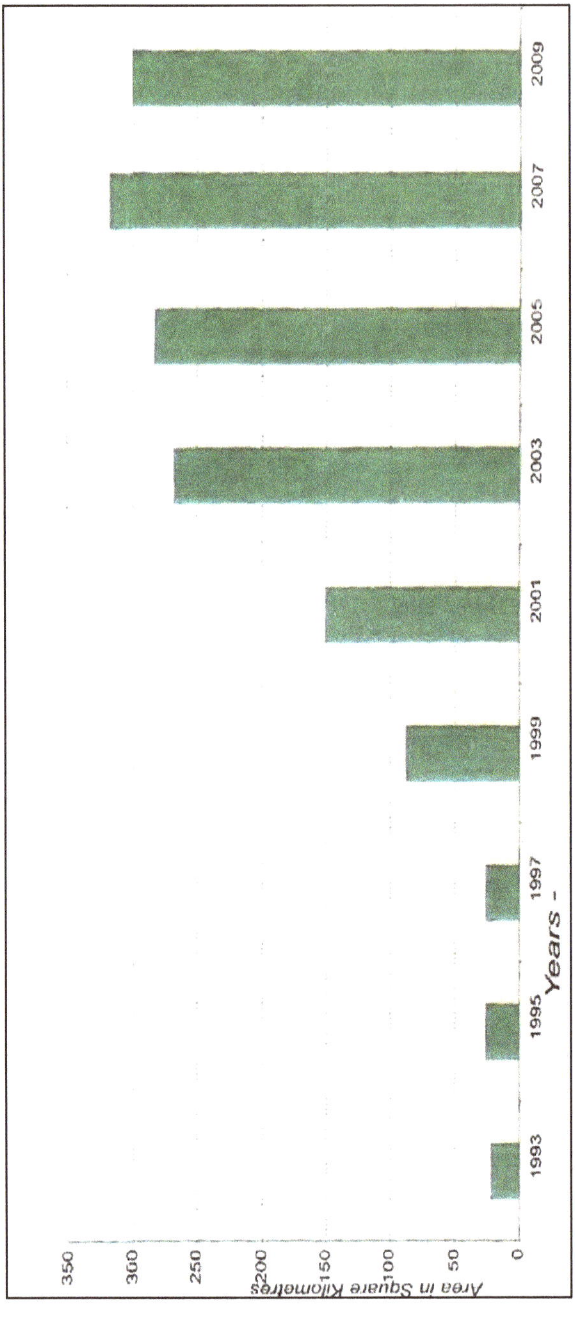

Figure 12.1: Change in Forest and Tree Cover in Delhi

Table 12.2: Change in Forest and Tree Cover in Delhi

Sl.No.	*Assessment Year*	*Forest and Tree Cover in Sq. km*	*Percentage of Geographical Area*
1.	1993	22	1.48
2.	1995	26	1.75
3.	1997	26	1.75
4.	1999	88	5.93
5.	2001	151	10.2
6.	2003	268	18.07
7.	2005	283	19.08
8.	2007	326	21.98
9.	2009	300	20.20

Source: State of Forest reports–1993, 1995, 1997, 1999, 2001, 2003, 2005, 2007 and 2009

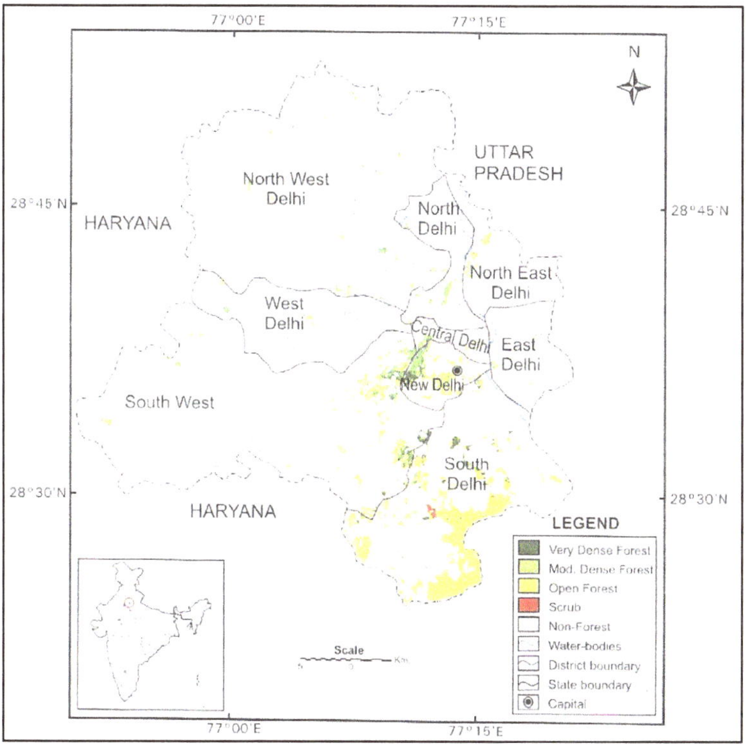

Figure 12.2: Forest Cover of Delhi

***Source*: State of Forest Report 2009, p. 75.**

peepal tree, for instance is estimated to give out 600 kg of oxygen in 24 hours. Being self renewing, the leaves act as persistent absorbers of pollutants, filtering particulates from the air. Forests acts as shelter belt or Buffer Zones against the noise and air pollution in Delhi.

Delhi forests provides increasingly polluted capital with oxygen as well as absorption of pollutants, forests acts as massive sound insulator absorbing the hubbubs of the city and finally trees blocks the dust from the hot dusty *'Loo'* that sweeps the capital during summer. The green and leafy portion of the trees and plants have the capacity to filter dust, smoke and pollutants in the air. Multi layered forests can regulate the gases in the atmosphere and can effectively remove various gases such as sulphur without any adverse effects. Trees also reduce thermal pollution by checking heat, build up in the atmosphere by preventing the radiation of heat from base surface. The trees therefore, serve the purpose of significant pollutants sink for various pollutants. When trees absorbs harmful gases, they gradually turn these harmful gases pollutant into harmless metabolities through synthetics and other metabolic activities. The luxuriant levels and branches serve to restrain the harmful particles in the air from spreading cultivation of resistant trees represents the greatest possibilities for reducing injury caused to the environment by air pollutants. A plant can remove significant quantities of pollutants from the air without sustaining damage or retardation.

The replacement of narrow-leaved trees of the desert vegetation with the broad-leaved trees in Asola Forest will make better carbon sink. They can absorb more pollution and release oxygen, add to the green cover of the city of Delhi and lead to a rise in the water table. In fact the Asola needs to be replicated in the other parts of Delhi. The ridge is the only barriers between the desert of Rajasthan and Delhi.

Conservation and Mangement of Forests

Realising the importance of forests the Govt. of NCT has accorded highest priority towards conservation and management of Delhi forests. The unique step taken by Delhi Government towards conservation of the existing trees by enacting the *Delhi Tree Prevention Act, 1994,* Wherein permission for cutting of every tree, required to be inevitable cut, is given on the condition of *planting at least Ten Number of trees* whose implementation has already had a salutary effect in protecting trees because of imposition of heavy penalties on individuals, institutions, prestigious club, educational and religious institution etc. found involved in illegal tree felling.

The Delhi Government is making tremendous efforts through various government and semi government organization to increase the vegetative and forest cover of Delhi. The Govt. of NCT Delhi had constituted a *'Task Force'* comprising of various government organization local bodies and NGO's on whose recommendations the *Greening Delhi Campaign* is being launched every year since 1997-98. The *Greening Action Plan* is an attempt to co-ordinate the activities of various agencies involved in conservation management and expansion of tree cover in mega city. Around 18.85 lakh seedling of trees and shrubs were planted on public and private land in the year 2008-09.

Some of the steps adopted by the Govt. of N.C.T. of Delhi towards conservation and management of forests and tree cover are as below:

1. *Massive awareness campaign* launched through Radio, T.V., Newspapers and posters.
2. Launching of *'Plantation Drive'* in residential colonies and schools in association with Resident Welfare Associations (RWA's) and free distribution of sapling and seedling.
3. *Adopting Transplantation of Trees Policy*–More than 1000 trees have been transplanted by various agencies like Delhi Jal Board (DJB), Delhi Metro Rail Corporation (DMRC), Public Works Deptt. (P.W.D.) etc. There is a overall survival rate of 50 per cent in transplantation.
4. *Adopting Replacement Policy for trees*–It is essential to create a second crop of trees through under-planting under mature trees well in advance.
5. *Rehabilitation of Bhatti Sanctuary through Eco-Task Force*–over 1,40,000 trees were planted and seeds of various indigenous trees species were sown at various location. Thirty check dams were constructed to conserve groundwater in the year 2006-2007.
6. *Enrichment of vegetation in existing forests area*–Efforts are being made to plant natural herby species of Delhi region *e.g., Dhak, Neem, Peepal, Amla, Jamun, Gular Desi Kikar* etc.
7. *Ridge Management Board* set up under the Chairmanship of Chief Secretary to develop ridge into a multilayered,

multi-storeyed forest. On the orders of the Honorable Supreme Court of India, many areas have been retrieved from encroachment, such as educational and religious institutions etc.

8. *Massive plantation* done along the railway tracks, drains, roadside, Goan Sabha lands etc.
9. *Augmentation of natural regeneration* in the Southern Ridge, especially in Sanctuary habitat.
10. Involvement of school children, *formation of Eco-club in school*, distribution of free sapling among the children and creating environment awareness in them.

Recommendations

1. There should not be any bare space at the traffic intersections. It should be covered either by grass or bush type species so that dust pollution during summer can be minimized.
2. Avenues of species are recommended on the outer sides of the traffic intersections. An outer row of shady hardy, indigenous pollution tolerant tree species and an inner row of ornamental flowering trees or shrubs are desirable. The outer row should be composed of evergreen shady trees with dense foliage and in between trees suitable under shrubs should be planted.
3. A macro level landscape plan is suggested. A minimum of 100 m wide green belt should be created around the borders of Delhi to filter out dust and heat waves.
4. For re-plantation process, emphasis should be given to the vehicular pollution tolerant indigenous species.
5. Before preparing project, environmental and forest aspect should be properly studied.

Conclusion

Green Zones are the key components to the well being of Delhi because trees play an important role in both reducing the environmental pollution load as well as serve as pollution indicators. The most effective means of combating high pollution level and regularly degrading environment of Delhi is though a well planned afforestation and *Greening Programme*. There has been substantial increase in forest and tree cover from 22 sq. km (1993) to 326sq. km (2007) which is 21.98 per cent of the geographical area of Delhi, but in 2009 it has declined to 299.58 sq. km which is 20.20 per cent of geographical area of Delhi. This decline can be attributed to massive cutting of trees by Delhi Metro Rail Corporation (DMRC). The multi-fold efforts for achieving this greening mission were-development of city forests, protection of the Delhi ridge, eco-restoration of degraded Bhatti mines area, development of Asola Wildlife Sanctuary, protection and maintenance of existing green belt, implementation of compensatory plantation schemes, greening of parks and gardens, roads, central verges, traffic roundabouts and conservation and restoration of water bodies and greening around them.

However the distribution of forest in Delhi is not uniform. More lot is to be done as per the *National Forest Policy 1988* which has set a goal that 33 per cent of the geographical area should be under forest and tree cover. The sustainable management of forest resource is imperative for ensuring ecological security in Delhi.

References

Champion, H.G. et al., 1968. *A Revised Survey of the Forest Type of India.* Govt. of India, New Delhi.

Chattopadhyay, S. and Prasad, N., 2006. *Essays on Water*. IRIS Publication Pvt. Ltd., New Delhi, p. 287.

Cunningham, William P. and Cunningham, Mary Ann, 2007. *Principles of Environmental Science.* Tata McGraw-Hill Publishing Company Ltd., New Delhi, p. 128.

Greening Delhi, Action Plan, 2007–2008. Published by Deptt. of Forest and Wildife, Govt. of Delhi.

India State of Forest Report, 2005, 2007 and 2009. Forest Survey of India, Dehradun.

Kalpavrakish, 1991. *The Delhi Ridge Forest: Decline and Conservation*, pp. 1–49.

Lavraj Kumar Committee Report, 1993. To recommend the pattern of management of Delhi Ridge.

Maheshwari, J.K., 1963. *The Flora of Delhi*. CSIR, New Delhi.

Mahto, Y. and Banotar, R.K., 1977. Bio-ecological studies of acridids of Delhi Region–II: Geographic distribution of genera and species. *Bull. Ent.*, 18: 25–45.

Towards Cleaner Air: A Case study of Delhi, 2003 Published by Deptt. of Environment and Forests Govt. of NCT of Delhi.

Chapter 13

Geographical Appraisal of Water Supply of a Metropolitan City: A Case Study of Delhi

B.S. Rathore[1] *and Chitra Sejwar*[2]

[1]*Head, Department of Geography, Government P.G. College, Morena, M.P.*

[2]*Assistant Professor, J.S. (P.G.) College, Sikandarabad, C.C.S. University, Meerut, U.P.*

Introduction

The population of Delhi has increased rapidly in last 10 decades. From a figure of 4,00,000 in year 1901, the population of Delhi has increased to around 1,67,53,235 in year 2011. Since Delhi acts as a magnet to attract migrants, estimated figures say that 2,00,000 to 3,00,000 people a year settle in Delhi permanently from other states. The population of Delhi is expected to rise 40 per cent by the year 2020. The continuously increasing demand for water has put the total water supply system of N.C.T. of Delhi under great strain. The demand of 1140 mgd (million gallon daily) has outstripped the present supply which is about 850 mgd. There is thus a need to augment the water availability to release the pressure on the strained water supply system of Delhi. The Delhi Jal Board (D.J.B.) is responsible for production of water in Delhi.

In Delhi the source of surface raw water include the river Yamuna that supplies 65 per cent, the river Ganga which contributes 25 per cent and another 10 per cent is supplied from groundwater through Ranney wells constructed in flood plains of Yamuna.

Study Area

Delhi lies between 28°24′17" and 28°30′0" North latitude and 76°50′24" and 77°20′37" East longitudes. It is surrounded on three sides, *i.e.*, on north, west and south by Haryana and to the east across the river Yamuna by Uttar Pradesh. The major part of Delhi lies on the western side of river Yamuna, while only some villages and urban areas of north-east and eastern districts lies on the eastern side of the river. Its greatest length is about 51.90 km and the greatest breath is 48.48 km. Delhi's altitude range between 213 and 305 metre above mean sea level. It covers 1,483 sq.km area (558.32 sq.km rural area and 924.68 sq. km urban area). According to 2011 census the total population of Delhi was 1,67,53,235 persons. The geographical area of Delhi is like a board alluvial neck, bounded by Thar Desert on the south, by mighty Himalayas on the north, Indo-gangetic plains on the north east, the fertile gangetic plains on the east and by Aravalli hills on the south. The topography of Delhi has two important physical features- Ridge and river Yamuna. The average annual rainfall ranges between 400-600mm.The average annual temperature ranges between 22° to 45°celsius.

Objectives

The objectives of the instant research paper are as follows:

- ☆ To study spatio-temporal changes of population and water supply in Delhi.
- ☆ To study the role of Delhi Jal Board in supplying potable water in Delhi.
- ☆ To assess the water conservation and management practices in Delhi.
- ☆ To assess the future sources of water.

Methodology

The study is based on secondary data collected from Census of India, data from Delhi Jal Board and other govt. agencies. The present work strives to bring to light the spatial variation distribution of water. Facts are arranged in a systematic order, mapped, described and interpreted in tables and bar diagrams.

Water Supply Scenario

The Population of Delhi is growing at a rapid rate in last 20 years. The total population growth in this decade was 20.96 per cent while in previous decade it was 46.31 per cent. The population of Delhi forms 1.38 per cent of India in 2011. In 2001, the figure was 1.35 per cent. Migration is the most vital factor and plays an important role in growth of population in Delhi.

Table 13.1: Growth of and Availability of Water in Delhi

Census Year	*Population*	*Growth (Per cent)*	*Water Produced M.G.D.*	*Availability of Water Gallon per Capita Daily (GPCD)*
1951	1744072	–	42	24
1961	2658612	52.4	91	37
1971	4065698	52.9	177	44
1981	6220406	53.0	296	51
1991	9420644	51.4	469	51
2001	13782976	46.3	650	50
2011	16753235	21.6	860	46

The allocation for has not been sufficient to meet the raw water requirements. Although Delhi is situated on the banks of river Yamuna, it can draw water only as per specified allocation. The Western Yamuna canal and Upper Ganga Canal are the life lines bringing raw water to Delhi. The govt. of U.P. and Haryana divert the entire flow into their canals upstream Tajewala. To finalise Delhi's share in Yamuna waters, the matter was taken by the Union Minister for water resources with the chief ministers of basic states *viz.* U.P., Haryana, Rajasthan, Delhi and Himachal Pradesh. Finally a Memorandum of understanding was signed on 12.5.94 for sharing of Yamuna water as under:

Table 13.2: Sharing of Yamuna Water by Different States

Sl.No.	*State*	*Water Sharing (B.C.M)*
1	Haryana	5.730
2	Uttar Pradesh	4.032
3	Rajasthan	1.119
4	Himachal Pradesh	0.378
5	Delhi	0.724
	Total	11.983

Source: DJB, Head Office, Varunalaya Building, New Delhi.

Spatial Inequity in Water Supply

The water supply in Delhi is far from uniformly distributed. The New Delhi Municipal Council (NDMC) area/Delhi cantonment area gets average supply above 450 litres per capita per day (LPCD) while Mehrauli gets 29 LPCD. The map below shows the spatial inequity in water supply in various ones of Delhi.

The latest water supply manual published by the Central Public Health and Environmental Engineering Organisation, Ministry of Urban Development and Poverty Alleviation, Government of India, recommends water requirements for domestic as 15 lpcd 15 per cent losses *i.e.* 172 LPCD. Considering requirement for other uses, the total demand works out to 274 LPCD.

From the Figure 13.1 it can be seen that except New Delhi Municipal Council (NDMC), Cantonment area, Karol Bagh and Civil Lines and Rohini zones. The rest of the city has water problem *i.e.* low water supply, erratic water supply *i.e.* 5 zones out of 12 zones (41.66 per cent) are getting water supply as per the norms of 274 lpcd and above.

The future raw water supplies for Delhi are proposed to be augmented through the following sources:

1. Bhakra Storage
2. Sub-Surface Water
3. Tehri Dam

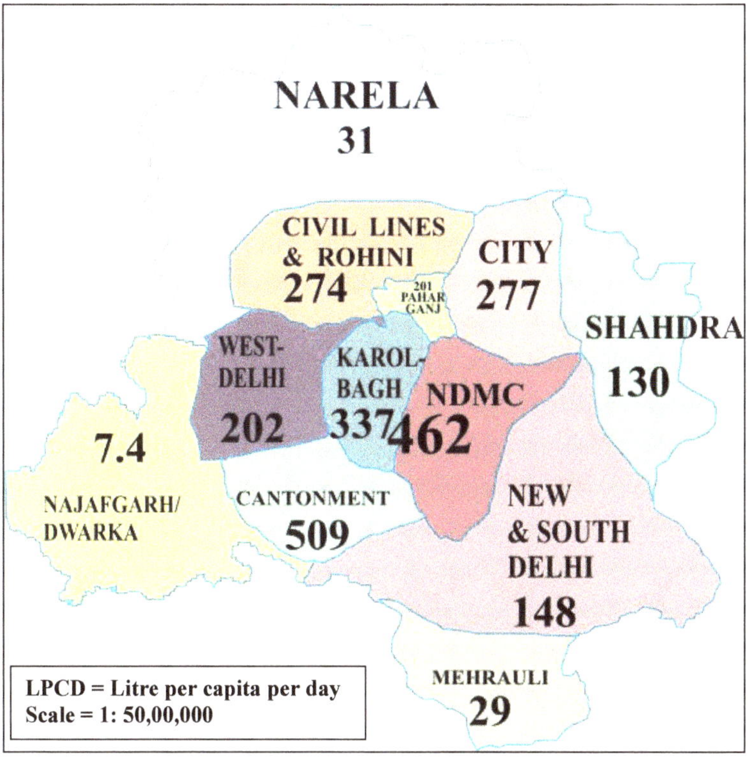

Figure 13.1: Spatial Inequity in Water Supply Litres per Capita per day (LPCD) in Various Zones of Delhi

***Source*: Delhi 1999: A fact, NCR Planning Board**

4. Renuka Dam
5. Kishau Dam
6. Lakhwar Dam

Source: WAPCOS

Measures taken to Conserve Water

1. To minimize conveyance losses for raw water, construction of 102 kms. Parallel lined channel has been taken up recently at a cost of Rs. 315 from Munak to Haiderpur.

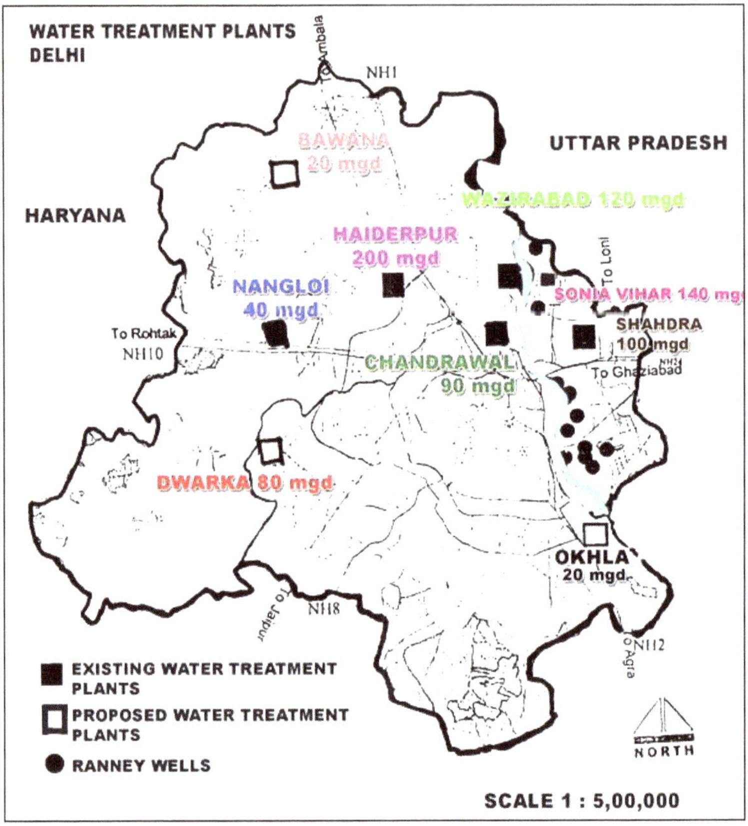

Figure 13.2: Future Water Supply to Delhi

2. A saving of 100 mgd of raw water on account of reduction of seepage losses in canal in U.P. by their lining at a cost of Rs. 1200 crorcs.
3. Rain water harvesting system has been made mandatory for all large building. D.J.B. has provided Roof top harvesting arrangements in 80 of its buildings covering an area of about 9 lacs sq.m Which can yield about 110 mgd annually.
4. Replacement of aged distribution pipe network is being done to prevent leakages and contamination in distribution system. 1000 kms replaced in last four years.

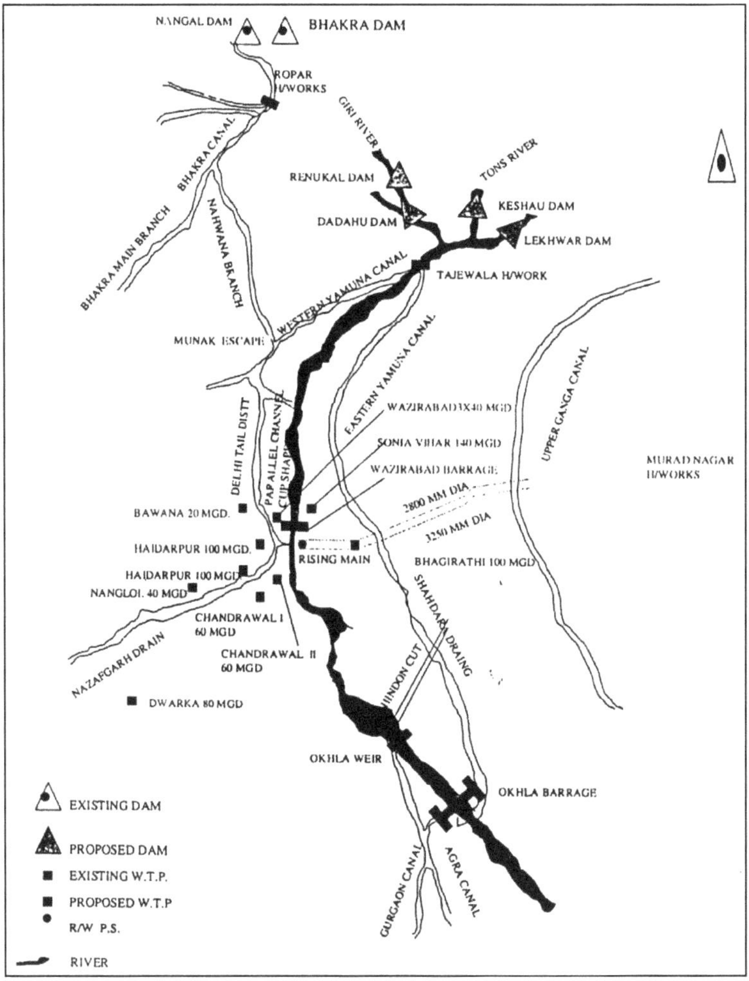

Figure 13.3: Future Sources of Raw Water for Delhi

5. By intensive monitoring about 11500 leakages have been detected and repaired during last six months.

Long Terms Measures for Supply of Raw Water

Parallel Lined Canal

Delhi's Haiderpur water treatment plant receives about 425 cusec (cubic feet per second) raw water through the Western Yamuna

Table 13.3: Proposed water availability/installed capacity

Sl.No.	Sources of Raw Water	Water Treatment Plants	Capacity (mgd) 2011	Capacity (mgd) 2021
1.	River Yamuna	Chandrawal I and II	90	90
2.	River Yamuna	Wazirabad I, II and III	120	120
3.	Bhakra Storage/Yamuna	Haiderpur I& II	200	200
4.	Bhakra Storage	Nangloi	40	40
5.	Upper Ganga Canal	Bhagirathi	100	100
6.	Upper Ganga Canal	Sonia Vihar	140	140
7.	Augmentation by recycling of Back wash water	Bhagirathi, Chandrawal Wazirabad, Haiderpur	46	46
8.	Groundwater	R/wells, T/wells.	103	103
9.	Saving in conveyance losses	Bawana Water works	20	20
	Plus additional resources	Dwarka	20	40
		Okhla Water works	20	20
		Total	899 against 1140	919 against 1380

Source: DJB, Head Office, Varunalaya Building, New Delhi

Canal System. The enroute losses in the 102 km stretch between Munak and the plant are estimated to be about 30 per cent at present. A lined channel running parallel to the existing one is being constructed by Haryana at Delhi's cost The work is estimated to cost Rs. 385 crore. With the constructions of this channel, about 160 cusec water will be additionally available to Delhi as a result of saving on conveyance losses enabling Delhi to set up additional water treatment plants.

Tehri Dam

300 Cusec water has been allocated to Delhi, out of the storage of the Tehri Dam. Water is to be conveyed through the Upper Ganga Canal. Conveyance losses from Haridwar to Murad Nagar have been assessed as 30 cusec. Therefore, 270 cusec raw water will be available at Murad Nagar. The Delhi Jal Board has taken up construction of 140-MGD water treatment plant at Sonia Vihar in Trans-Yamuna Area at a cost of Rs. 188 crore. The work is in progress and is targeted for completion by the end of 2003. To bring raw water for the plant, construction of RCC onduit from Murad Nagar to Sonia Vihar has been awarded by the UP Jal Nigam on behalf of the Delhi Jal Board to NBCC at a cost of Rs. 87 crore.

Kishu Dam

372 MGD raw water has been earmarked for Delhi's use in the storage of this dam. This dam is to be constructed on the river Tons in Uttarakhand.

The Western Yamuna Canal is one of the life-lines bringing raw water to Delhi. In case of breakdown of the canal, to provide connectivity to the two existing water treatment plants at Haiderpur and Wazirabad, a 2000 mm raw water main in a length of 11 km from Waziarabad to Haiderpur will be laid so as to provide alternative raw water arrangement. The work is proposed to be taken up shortly.

In order to exploit groundwater in flood plains, 100 tubewells will be constructed along the Right Marginal Embankment in Alipur Block in Pali/Burari areas. 40 of these tubewells have already been commissioned. The work is in progress and will yield about 30 MGD water. Work has also commenced on tube wells near Kalindi.

Suggestions

1. The National capital Region (NCR) should be developed rapidly if the Delhiites are to get reasonable supply of water everyday.
2. The effective demand side management regime be put to use.
3. The artificial recharge projects should be taken up all over Delhi to check decline in water levels.
4. Conservation and management of water is the need of the hour. Water saved is water produced.
5. Rain Water Harvesting (RWH) is a vialable technology can become a integral habit.
6. Public awareness campaign should be launched on warfooting to educate public to use water judiciously.
7. Conservation of water should be the responsibility of the *individual* and the *nation* as a whole.
8. The groundwater resources though being developed extensively, hold further scope for augmentation of water supplies.

Conclusion

Delhi is in the grip of massive water crisis. The continuously increasing demand for water has put the total water supply system of N.C.T. of Delhi under great strain.

Management of water in Delhi needs to be made more effective and innovative. Non-structural approaches in water management will have to replace the large scale supply-side approaches of the past.

Delhi Jal Board (D.J.B.) will have to take recourse to a number of innovative schemes like recycling of wastewater at water treatment plants, optimization of water production and exploitation of groundwater in the flood plains of the river-Yamuna, rain water harvesting and conservation of water.

There is a wide gap between demand and supply to different types of settlements which is against the concept of social equity. The problem faced by Delhiiites due to spatial inequity in water supply can be over come by rationalizing the distribution system and increasing the institutional efficiency by better management.

References

Central Public Health Environment Engineering Organization Manual, Ministry of Urban Development and Poverty Alleviation, Government of India.

Delhi Jal Board (D.J.B.) Head Office, Varunalaya Bldg. New Delhi–110005

Government of the National Capital Territory of Delhi, 2010. Delhi Statistical Handbook, Directorate of Economics and Statistics.

Kumar, Ashok, 2004. Management of safe drinking water in Delhi. In: *Water Resources Management*, (Eds.) G.S. Chouhan and R.N. Dubey. Shri Natraj Prakashan, p. 34–49.

Majumdar, Paramita, 2004. Challenges of water management in India. In: *Resource Conservation and Food Security*, (Ed.) Tapeshwar Singh. Concept Publishing Company, New Delhi, 1: 76.

Proceeding of the Seminar: *Water Conservation: "Sustainable Solutions to Save Delhi from Water Crisis"* organised by Delhi Jal Board (D.J.B.) and Confederation of Indian Industry (CII) in May 2003.

Sinha, Saurabh, 2002. Delhi dithers as water table plunge alarmingly. *The Times of India*, New Delhi.

Status Report for Delhi 21, 2010. *Delhi Urban Environment and Infrastructure Improvement, Projects (DUEIIP)*, Government of India, Ministry of Environment and Forests and Government of NCT of Delhi, Planning Department.

Swamy, S.A., 1987. Delhi water supply schemes: Issues, problems and solutions. In: *A Paper Prepared for Training Workshop on Urban Civil Services and Management*, Organized by TCPO, New Delhi.

Chapter 14

Climate Change and its Effects on Insects, Pests, Pathogens and Agriculture Environment

Swati Singh, Y.P. Singh, Dharvendra Singh, K.K. Yadav, Reeta Mishra and S.P.S. Rajput

RVSKVV, Krishi Vigyan Kendra, Morena, M.P.

The three legs of the triangle–host, pathogen, and environment–must be present and interact appropriately for plant disease to result. If any of the 3 factors is altered, changes in the progression of a disease epidemic can occur. The major predicted results of climate change–increases in temperature, moisture and CO_2–can impact all three legs of the plant disease triangle in various ways.

How Rising Temperatures will Affect Pathogens and Disease

Temperature has potential impacts on plant disease through both the host crop plant and the pathogen. Research has shown that host plants such as wheat and oats become more susceptible to rust diseases with increased temperature; but some forage species become more resistant to fungi with increased temperature.Many mathematical models that have been useful for forecasting plant disease epidemics are based on increases in pathogen growth and infection within specified temperature ranges. Generally, fungi that cause plant disease grow best in moderate temperature ranges. Temperate climate zones that include seasons with cold average temperatures are likely to experience longer periods of temperatures suitable for pathogen growth and reproduction if climates warm. For example, predictive models for potato and tomato late blight (caused by *Phytophthora infestans*) show that the fungus infects and reproduces most successfully during periods of high moisture that occur when temperatures are between 45°F (7.2°C) and 80°F (26.8°C) Earlier onset of warm temperatures could result in an earlier threat from late blight with the potential for more severe epidemics and increases in the number of fungicide applications needed for control.

How Changes in Moisture will Affect Pathogens and Disease

Moisture can impact both host plants and pathogen organisms in various ways. Some pathogens such as apple scab, late blight, and several vegetable root pathogens are more likely to infect plants with increased moisture–forecast models for these diseases are based on leaf wetness, relative humidity and precipitation measurements. Other pathogens like the powdery mildew species tend to thrive in conditions with lower (but not low) moisture. More frequent and extreme precipitation events that are predicted by some climate change. Models could result in more and longer periods with favourable pathogen environments. Host crops with canopy size limited by lack of moisture might no longer be so limited and may produce canopies that hold moisture in the form of leaf wetness or high canopy relative humidity for longer periods, thus increasing the risk from pathogen infection Some climate change models predict higher atmospheric water vapor concentrations with increased temperature–this also would favour pathogen and disease development.

How Rising CO_2 Levels will Affect Pathogens and Disease

Increased CO_2 levels can impact both the host and the pathogen in multiple ways. Some of the observed CO_2 effects on disease may counteract others. Researchers have shown that higher growth rates of leaves and stems observed for plants grown under high CO_2 concentrations may result in denser canopies with higher humidity that favor pathogens. Lower plant decomposition rates observed in high CO_2 situations could increase the crop residue on which disease organisms can overwinter, resulting in higher inoculums levels at the beginning of the growing season, and earlier and faster disease epidemics. Pathogen growth can be affected by higher CO_2 concentrations resulting in greater fungal spore production. However, increased CO_2 can result in physiological changes to the host plant that can increase host resistance to pathogens.

How Climate Change could Impact Plant Disease Management Practices

While physiological changes in host plants may result in higher disease resistance under climate change scenarios, host resistance to disease may be overcome more quickly by more rapid disease cycles, resulting in a greater chance of pathogens evolving to overcome host plant resistance. Fungicide and bactericide efficacy may change with increased CO_2, moisture, and temperature. The more frequent rainfall events predicted by climate change models could result in farmers finding it difficult to keep residues of contact fungicides on plants, triggering more frequent applications. Systemic fungicides could be affected negatively by physiological changes that slow uptake rates, such as smaller stomatal opening or thicker epicuticular waxes in crop plants grown under higher temperatures. These same fungicides could be affected positively by increased plant metabolic rates that could increase fungicide uptake. It is not well understood how naturally-occurring biological control of pathogens by other microbial organisms could change as populations of micro-organisms shift under changed temperature and moisture regimes– in some cases antagonistic organisms may out-compete pathogens while in

Others pathogens may be favoured. Exclusion of pathogens and quarantines through regulatory means may become more difficult for authorities as unexpected pathogens might appear more frequently on imported crops.

How this will Affect Farmers

Although the specific impacts of climate change on plant disease are difficult to predict given our current knowledge, it seems possible to make several generalizations for farmers. Increased winter temperatures will likely mean higher populations of pathogens survive to initially infect plants. Increased temperatures will likely result in northward expansion of the range of some diseases because of earlier appearance and more generations of pathogens per season more frequent and more intense rainfall events will tend to favour some types of pathogens over others two pathogens important in the northeastern US, Stewart's wilt and late blight, illustrate some of these effects. It can be a devastating disease for both crops and farmers, with complete crop loss a possibility if control measures are not employed. Infection is triggered by high moisture conditions within a fairly specific temperature range. Annually, 5-20 fungicide applications from as early as June through August.

Relation with Insects and Environment

Insects are cold-blooded organisms–the temperature of their bodies is approximately the same as that of the environment. Therefore, temperature is probably the single most important environmental factor influencing insect behaviour, distribution, development, survival, and reproduction. Insect life stage predictions are most often calculated using accumulated degree days from a base temperature and bio fix point. Some researchers believe that the effect of temperature on insects largely overwhelms the effects of other environmental factors It has been estimated that with a 2°C temperature increase insects might experience one to five additional life cycles per season. Other researchers have found that moisture and CO_2 effects on insects can be potentially important considerations in a global climate change setting.

How Rising Temperatures Affects Insects

Climate change resulting in increased temperature could impact crop pest insect populations in several complex ways. Although some climate change temperature effects might tend to depress insect populations, most researchers seem to agree that warmer temperatures in temperate climates will result in more types and higher populations of insects. Increased temperature could increase pest insect populations.

Researchers have shown that increased temperatures can potentially affect insect survival, development, geographic range, and population size. Temperature can impact insect physiology and development directly or indirectly through the physiology or existence of hosts. Depending on the development "strategy" of an insect species, temperature can exert different effects some insects take several years to complete one life cycle–these insects (cicadas, arctic moths) will tend to moderate temperature variability over the course of their life history. Some crop pests are "stop and go" developers in relation to temperature–they develop more rapidly during periods of time with suitable temperatures. We often use degree-day or penology based models to predict the emergence of these insects and their potential to damage crops (*cabbage maggot, onion maggot, European corn borer, Colorado potato beetle*). Increased temperatures will accelerate the development of these types of insects–possibly resulting in more generations (and crop damage) per year.

"Migratory" insects (corn earworm in northern parts of the northeast) may arrive in the North-east earlier, or the area in which they are able to over winter may be expanded. Natural enemy and host insect populations may respond differently to changes in temperature. Parasitism could be reduced if host populations emerge and pass through vulnerable life stages before parasitoids emerge. Hosts may pass though vulnerable life stages more quickly at higher temperatures, reducing the window of opportunity for parasitism. Temperature may change gender ratios of some pest species such as trips potentially affecting reproduction rates. Insects that spend important parts of their life histories in the soil may be more gradually affected by temperature changes than those that are above ground simply because soil provides an insulating medium that will tend to buffer temperature changes more than the air Lower winter mortality of insects due to warmer winter temperatures could be important in increasing insect populations. Higher average temperature might result in some crops being able to be grown in regions further north–it is likely that at least some of the insect pests of those crops will follow the expanded crop areas. Insect species diversity per area tends to decrease with higher latitude and altitude meaning that rising temperatures could result in more insect species attacking more hosts in temperate climates. Based on evidence developed by studying the fossil record some researchers conclude that the diversity of insect species and the intensity of their feeding have increased historically with increasing temperature.

Increased temperature could decrease pest insect populations some insects are closely tied to a specific set of host crops. Temperature increases that cause farmers not to grow the host crop any longer would decrease the populations of insect pests specific to those crops. The same environmental factors that impact pest insects can impact. Their insect predators and parasites as well as the disease organisms that infect the pests, resulting in increased attack on insect populations. At higher temperatures, aphids have been shown to be less responsive to the aphid alarm pheromone they release when under attack by insect predators and parasitoids–resulting in the potential for greater predation.

Summary

- ✰ The precise impacts of climate change on insects and pathogens is somewhat uncertain because some climate changes may favor pathogens and insects while others may inhibit a few insects and pathogens. The preponderance of evidence indicates that there will be an overall increase in the number of outbreaks of a wider variety of insects and pathogens.
- ✰ The possible increased use of fungicides and insecticides resulting from an increase in pest outbreaks will likely have negative environmental and economic impacts for agriculture.
- ✰ The best economic strategy for farmers to follow is to use integrated pest management.

Practices to closely monitor insect and disease occurrence. Keeping pest and crop management records over time will allow farmers to evaluate the economics and environmental impact of pest control and determine the feasibility of using certain pest management strategies or growing particular crops.

Chapter 15

The Concept of Sustainable Development: Its Needs and Importance

*Satish Modi**

Assistant Professor of Commerce,

I.G.N. Tribal University, Amarkantak, M.P.

ABSTRACT

Development has traditionally been linked with depletion of the earth's natural resources and is now placing growing pressure on the earth's assimilative capacity. A crucial challenge confronting mankind today is how to meet the basic needs and requirement of all people without simultaneously destroying the resource base that is the environment, from which ultimately these needs have to be met. We can meet out this challenge through the sustainable development. The world commission on environment and development of United Nation defined sustainable development as the "development that meets the needs of the present without compromising the ability of future generations to meet their own needs." Thus sustainable development means development without destruction. It integrates environment and development. It is

* Author E-mail: smodiy2k@yahoo.com

the creation of sustainable improvements in the quality of life of all people through increase in real income per capita, improvement in education, health and general quality of life and improvements in quality of natural environmental resources. In other words, it is a situation in which economic development does not decrease over time. It can be modified as a path of development in which options of future generations are not compromised by the path taken by the present generation. Sustainable Development is development that is everlasting. Today we have to enter sustainability in development in all our planning process as one of the basic and permanent objectives. We have to balance in economic goal and ecology goal. Economic goal will definitely damage the ecological goal as a common perception, but through proper management both could be compatible to each other. Today economic planning and environmental protection have identical goals of sustainable development. Therefore both must be deeply integrated in the planning process. The sustainability must be strengthened as a guiding principle of development.

Introduction

Development and environment has inverse relation. Development has traditionally been linked with depletion of the earth's natural resources and is now placing growing pressure on the earth's assimilative capacity. A crucial challenge confronting mankind today is how to meet the basic needs and requirement of all people without simultaneously destroying the resource base that is the environment, from which ultimately these needs have to be met. We can meet out this challenge through the sustainable development. The term 'Sustainable Development' was first used by the 'World Conservation Strategy' presented by the International Union for the Nature and Natural Resources in 1980.[1] The world commission on environment and development of United Nation defined sustainable development as the "development that meets the needs of the present without compromising the ability of future generations to meet their own needs."[2] Thus sustainable development means development without destruction. It integrates environment and development. It is the creation of sustainable improvements in the quality of life of all people through increase in real income per capita, improvement in education, health and general quality of life and improvements in quality of natural environmental resources. In other words, it is a situation in which economic development does not decrease over time. It can be modified as a path of development in which options of future generations are not compromised by the path taken by the present generation. Sustainable Development is development that is everlasting. It contributes to the quality of life through improvements in natural environments. In turn, it supply utility to individuals, inputs to the process of economic and service that support human life.

According to Pearce and Warford "Sustainable Development is a process in which natural resource base is not allowed to deteriorate. It emphasizes the hitherto unappreciated role of environmental quality and environmental inputs in the process of rising real income and quality of life."[3] The World Development Report of 2000 emphasizes the creation of sustainable improvements in the quality of life for all people as the principal goal of development policy. The main objectives of sustainable development are stated as under:

- Creation of sustainable improvements in the quality of life for all people.
- Increasing economic growth through meeting basic needs *i.e.*, raising the standard of living.
- Providing chance to participate in public life and helping to clean environment.
- Aiming at maximizing the net benefits of economic development subject to maintaining the stock of all environmental and natural resources assets.
- Aiming at accelerating economic development in order to conserve and enhance the stock of environmental, human and physical capital without making future generations worse off.
- Aiming to require strong sustainability that the natural capital stock should not decline. Moreover, weak sustainability needs that total value of physical, human and natural capital stock should not decrease.

The main reason people are concerned with sustainable development is because of the fact that the world is a place of limited materials. Sustainable development aims to establish a system of resource consumption that both meets the needs of human life and leaves the environment healthy enough to continue to produce the resources future generations will need. The scope of sustainable development is massive, and many areas of life can fall under this broad umbrella. The following are some of the most important issues to sustainable development.

Freshwater

This is one of the most essential issues facing the species since humans have walked on the earth. Of the planet's water, 97.5 per cent of it is salt water and of the 2.5 per cent that is fresh, 2 per cent of that is frozen in the polar ice caps.[4] This means all of the humans and animals on the world survive on less the 1 per cent of the planet's total water supply. The problem is that the population is increasing the demand for water while the supply is decreasing due to issues like pollution. In order to become more sustainable, humans need to take action. Agricultural systems must become more efficient worldwide and some subsidies on water that make it very inexpensive in certain areas should be lifted to allow the price to rise, which will encourage conservation.

Poverty

Poverty is said to be both cause and effect of environmental degradation. The circular link between poverty and environment is an extremely complex phenomenon. Inequality may foster unsustainability because the poor, who rely on natural resources more than the rich, deplete natural resources faster as they have no real prospects of gaining access to other types of resources. Moreover, degraded environment can accelerate the process of impoverishment, again because the poor depend directly on natural assets. In 1972, the Prime Minister of India, Mrs. Indira Gandhi emphasized, at the UN conference on 'Human Environment at Stockholm,' that the removal of poverty is an integral part of the goal of an environmental strategy for the world.[5]

Finance

One of the main issues of sustainable development is figuring out how to best distribute the country's wealth. In India out of 1.2 billion citizens only 53 citizens are dollar billionaires. The combined net worth of these 53 billionaires was $ 335 billion; equal to one-third of country's GDP. The top ten richest Indians have a total net worth of $144.7 billion, which is approx. 15 per cent of India's GDP.[6] In a free society, disparities are always going to exist, but the problem confronting the future is that these numbers are trending towards the wealthy owning more and more. A truly sustainable society would be interested in increasing all its citizens' access to resources.

Health

Health is inextricably linked to sustainable development. In areas where poverty is rampant and resources are not provided equally to all members of society, the individuals at the lower levels suffer tremendously in terms of health. Nutritious food is limited in these areas as well as access to adequate medicine and medical health professionals. When infectious diseases like AIDS are brought into the mix, this issue quickly becomes an epidemic.

Natural Resources

Protecting and managing the natural resources is very essential for economic and social development. The integration of agriculture with land and water management and with ecosystem conservation is essential for both environmental sustainability and agricultural production.

These issues are some of the most important issues to sustainable development, but other areas like agriculture, technology, sanitation, human settlements, biotechnology and many others also play a role.

Following are the important steps towards sustainable development.

- Population stabilization
- Integrated land use planning
- Maintenance of sustainability in agriculture through restoration of healthy cropland and grassland
- Conservation and rational exploitation of forest wealth and regeneration in wasteland and deforested areas
- Conservation of biological diversity
- Control of pollution of atmosphere (air) and hydrosphere (water)
- Development of non-polluting renewable energy systems

- ☆ Recycling of wastes and residues
- ☆ Imposition of ecologically compatible human settlements and slum improvement
- ☆ Updating environmental laws and its strict imposition
- ☆ Assessment of ecological security

Measures/Conditions for Sustainable Development

Generally, it is difficult to measure sustainable development. It involves the valuation of environmental damage and comparing it with the cost of preventing it. It faces many problems like measuring of capital stock, natural resource accounting, measuring environmental values and the use of an appropriate discount rate for maintaining and optional balance between the use and preservation of natural resources. However, here we discuss conditions for the attainment of sustainable development. These are following:

Measuring Natural Capital Stock

The stock of natural resources assets or environment assets includes "soil fertility, forests, fisheries, and the capacity to assimilate waste, oil, gas, coal, the ozone layer and biogeochemical cycles." The requisite condition for sustainable development is that the natural capital stock should be conserved and improved or should remain at least constant. This can be measured in terms of the cost-benefit analysis of changes in the natural capital stock. Similarly, when the atmosphere is kept clean, it is beneficial for all of us. So sustainability is referred as maintaining and improving natural capital assets.

Natural Resources or Green Accounting

Another measure or condition of sustainable development is green accounting. It allows computation of income for a nation by taking into account the economic damage and depletion in the natural resource base of a country. It is a measure of sustainable income level that can be secured without decreasing the stock of natural assets. Thus, it needs that adjustment should be made in the system of national income accounts in terms of stock of natural assets. The computation of gross national product (GNP) would be replaced by a measure of national output that includes the economic cost of degrading natural resources. It required producing goods and services directly and indirectly. GNP would include the stock of natural assets and the measure of sustainable development can be [NNP = GNP- Dn][7] where, NNP is sustainable national income and Dn is depreciation of the monetary value of natural assets in a year. But the computation of such a measure of sustainable income is very difficult. This happens especially in computation of monetary valuation of non-marketed natural assets and externalities.

Measuring Environmental Values

In measuring environmental values, another problem occurs during comparison with the cost of preventing it. Generally, it is believed that environmental degradation leads to costs and it also improves benefits on resource users. But, the problem is how to measure costs and benefits of environmental effects on the present and the future generations. For this purpose a rate of discount is needed for discounting all costs and benefits. However, it also creates a lot of confusion among economists in discounting environmental costs and benefits.

Social Discount Rate

It concerns comparing the benefits of environmental protection with the costs incurred on it. Economists have suggested four approaches for economic valuation of environmental damage in World Development Report of 1992. These are Market Prices, Costs of Replacement, Surrogate Markets and Surveys.[8]

From this discussion we can conclude that same acceptable measure is to be evolved to measure sustainable development.

Suggestions for Achieving Sustainable Development

Following guidelines are suggested for achieving sustainable development.

- ☆ The foremost policing are to reduce poverty, which degrades humans and the environment by forcing people to use resources, unsustainably to stay alive. Therefore such projects should be started which provide greater employment opportunities to the poor class.

- ☆ We should use energy more efficiently and not use high quality energy for tasks that require only moderate quality energy.
- ☆ We should reduce the input of matter and energy resources to prevent excessive depletion and degradation of planetary sources.
- ☆ We should use potentially renewable resources and use them no faster than the rate at which they are regenerated.
- ☆ To shift from exhaustible and potentially polluting fossil and nuclear fuels to less harmful renewable energy obtained directly or indirectly from the sun.
- ☆ We should not waste nonrenewable resources and use them no faster than the rate at which a renewable resource, used sustainable can be substituted.
- ☆ We should develop and propagation of eco-friendly technologies for sustainable development.
- ☆ We should emphasize pollution prevention and waste reduction instead of pollution clean up and waste management.
- ☆ We should control human population growth to help reduce stress on global life support.
- ☆ We should develop a good and sustainable environment through natural resources management.
- ☆ We should not deplete or degrade Earth's physical, chemical and biological capital, which supports all life and human economic activities.
- ☆ We should work with the rest of nature to sustain the ecological integrity, biodiversity and adaptability to earth's life support systems for us and other species.
- ☆ We should try to understand and cooperate with the rest of nature. The earth does not belong to us; we belong to the earth.
- ☆ We should evaluate the beneficial and harmful consequences of our lifestyle and profession on the earth, today and in the future.
- ☆ To create environmental consciousness through education and mass awareness programmed.

- To elicit and ensure participation of people in programmed for environmental improvement and for integrating the environmental concerns in planning and implementation of development programmed.
- To develop appropriate organizational structures and a pool of professional manpower to serve as the cadre for environmental management service.
- To develop people's willingness and ability to take up the task of saving the planet from environmental pollution is needed. Any of the environment policy, government legislation and regulation for environmental protection can be successful only with people's participation.
- Involving NGO's and women in all environmental related activities.
- We should abide by the ethical environmental standards of environmental association and professional societies to ensure that our environment is protected.

Conclusion

Developmental activities are always associated with environmental degradation. The development model followed in India so far does not confirm to the parameter of sustainable development because of materialistic thought there is heavy exploitation of natural resource. Hence today development is to be made along with preventing environment for future generation. We need to use natural resources like trustee not like owner, as we have to handover it to our future generation. Natural resources are not totally renewable and a serious as well as planned framework for sustainable development can help in their protection. Hence, today we have to enter sustainability in development in all our planning process as one of the basic and permanent objectives. We have to balance in economic goal and ecology goal. Economic goal will definitely damage the ecological goal as a common perception, but through proper management both could be compatible to each other. Today economic planning and environmental protection have identical goals of sustainable development. Therefore both must be deeply integrated in the planning process. The sustainability must be strengthened as a guiding principle of development.

References

(1) Bhushan, Bharat and Kumar, Vinit, 2006. *Environmental Management*. Shree Publishers and Distributors, New Delhi, p. 32.

(2) Chatwal, G.R. and Sharma, Harish, 2008. *A Textbook of Environmental Studies*. Himalaya Publishing House, Mumbai, p. 438.

(3) Mishra, R.M., 2009. *An Introduction to Environmental Science*. Shree Publishers and Distributors, New Delhi, p. 58.

(4) Murthy, D.B.N., 2006. *Environmental Planning and Management*. Deep and Deep Publications Pvt. Ltd., New Delhi, p. 19.

(5) Mishra, S.K. and Puri, V.K., 2008. *Indian Economy*. Himalaya Publishing House, Mumbai, p. 123.

(6) Datt, Ruddar and Sundharam, K.P.M., 2009. *Indian Economy*. S. Chand and Company Ltd., New Delhi, p. 460.

(7) Lekhi, R.K., 2009. *The Economics of Development and Planning*. Kalyani Publishers, New Delhi, p. 59/I.

(8) *Ibid*.

Chapter 16

Management of Soil Resources for Sustainable Agriculture

T.R. Rathore[1] *and Rajvir Singh Rathore*[2]
[1]*Ex. Professor of Soil Science, GBPUAT, Pantnagar*
[2]*Deputy Director General,*
U.P. Council of Agricultural Research, Lucknow

Introduction

The success of boosting up agricultural production has come through the high yielding varieties which were highly responsive to chemical fertilizers. Increased use of chemical fertilizers also promoted the large scale use of many other agro-chemicals *e.g.* fungicides, pesticides, herbicides in order to save the crops from the damages by diseases, insect pests, and others. The contribution of fertilizers in increasing agricultural production is well established. Further, the contribution of other critical inputs such as increased use of HYV/hybrid seeds, irrigation, pesticides in association with chemical fertilizers is also well recognised. However, the impact of these inputs in recent past years has not been significant as almost stabilization in the production and productivity of foodgrains is being noticed (Table 16.1).

Nutrient Status and Response to Applied Fertilizers in Indian Soils

Besides the stabilization in production and productivity, adverse effect of various agro-chemicals on soil health, biodiversity and other natural resources are also observed. Indian soils which were deficient only for nitrogen in year 1950-51 are showing multi-nutritional deficiency as depicted in Figure 16.1, due to unsystematic planning and over exploitation of soil resources. With passage of time, numbers of elements are being added in the list of nutritional deficiency in soils.

Table 16.1: All-India Area, Production and Yield of Foodgrains alongwith Coverage Under Irrigation

Year	*Area Million ha*	*Production Million Tonnes*	*Yield kg/ha*	*Area Under Irrigation (%)*
2001-02	122.78	212.85	1734	43.0
2002-03	113.86	174.77	1535	42.8
2003-04	123.45	213.19	1727	42.2
2004-05	120.00	198.36	1652	44.2
2005-06	121.60	208.60	1715	45.5
2006-07	123.71	217.28	1756	46.3
2007-08	124.07	230.78	1860	46.8
2008-09	122.83	234.47	1909	NA
2009-10	121.37	218.20	1798	NA

Note: The yield rates given above have been worked out on the basis of production and area figures taken in '000 units.

Source: Directorate of Economics and Statistics, Department of Agriculture and Cooperation.

Deficiency to different nutrients is not uniform in different part of the country as it varies from area to area *e.g.* 86 per cent soil samples of Maharashtra were deficient to zinc; whereas in Rajasthan only 21 per cent soil samples were found deficient. Similarly, boron deficiency was found highest in Orissa soils (69 per cent) as compared to Punjab soils (13 per cent). Over exploitation of soil resources have not only caused the multi-nutritional deficiency, but, the response to per unit fertilizer applied, specially in irrigated areas has gone down drastically. On an average the response to per kg NPK applied in irrigated conditions was around 13.4 kg grain during 1970, has now gone down to as low as 3.7 kg grain production in the year 2010 (Figure 16.2).

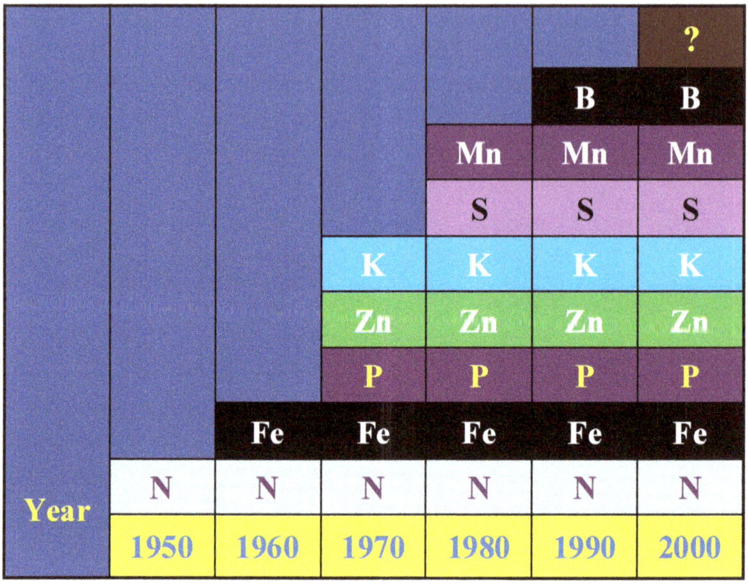

Figure 16.1: Emerging Multi-Nutritional Deficiencies in Soils

Therefore, it is clear that for higher yields on sustainable basis, other locally available resources and measures will have to be adopted for improving soil health and environment and enhancing

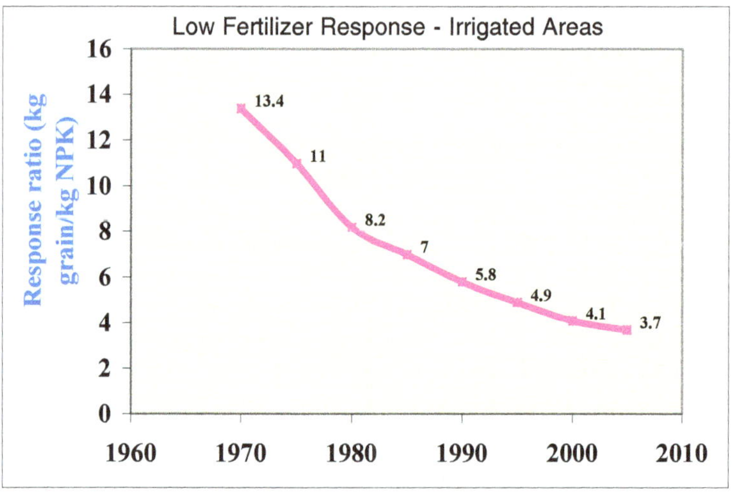

Figure 16.2: Declining Fertilizer Response

the agricultural production and productivity in order to meet the growing demand.

Land Use Pattern and Option

As per the norms of National Commission for Integrated Water Resource Development, foodgrains and feed demand for projected population of India *i.e.,* 137.01 and 162.29 crore by the end of the year 2025 and 2050 on the basis of medium growth rate of aggregate expenditure would be about 299 and 464 million tonnes respectively. The estimated future demand can be achieved either by horizontal growth *i.e.,* expansion in cultivable area or through vertical *i.e.,* increase in yield per unit area. As far the scope of horizontal expansion is concerned, it is almost negligible as the net cultivated area is nearly constant about 141 million hectares. It is also worthwhile to mention that a sizeable cultivable area annually is being diverted for non-agricultural uses, mainly for urbanization, industrilization and infrastructure development (Table 16.2).

Therefore, net sown area in the country (about 141 million hectare) is being maintained by bringing the waste lands (usar and uncultivable) under cultivation after their reclamation and improvement. The productivity of these newly developed lands is very low. On the other hand, we are unable to arrest good productive current fallow lands. The reasons of increasing current fallow lands need to be studied for taking necessary steps and formulating strategies for bringing these prim agricultural lands under crops. In India, of the geographical area of 329 million hectare as much as 130 million hectare area is suffering from various kinds of degradations. Various kinds of degraded lands are given in Table 16.3.

Also reduction in per capita cultivable lands from 0.34 hectare in 1950-51 to 0.12 hectare in 2010 is becoming less attractive for farming and also turning uneconomical. Therefore, under the discussed circumstances, without soil improvement it would be very difficult to maintain the present level of agricultural production and productivity in future. Therefore, only second option *i.e.,* vertical growth is available option and need to be exploited with utmost care to meet out the food grains and other demands of ever increasing population. The gap in potential yields recorded at research stations and actually realised at farmers field is very large (10 to 30 q/ha) and need to be bridged with the adoption of eco-friendly improved

Table 16.2: Agricultural Land Use in India (Million hectare)

Classification	1950-51	1960-61	1970-71	1980-81	1990-91	2000-01	2007-08
Geographical Area	328.73	328.73	328.73	328.73	328.73	328.73	328.73
Reporting Area for Land Utilisation Statistics (1 to 5)	298.46	303.75	304.16	304.86	305.18	305.12	305.67
1. Forest	54.05	63.83	67.46	67.81	69.53	69.41	69.63
Per cent	16.44	19.42	20.52	20.63	21.15	21.11	21.18
2. Not Available for Cultivation (A+B)	50.75	44.61	39.55	40.48	41.48	41.57	43.22
(A) Area Under Non-agricultural Uses	14.84	16.48	19.60	21.09	23.89	24.05	25.92
Per cent	4.51	5.01	5.96	6.41	7.27	7.32	7.89
(B) Barren and Un-culturable Land	35.91	28.13	19.96	19.39	17.59	17.52	17.29
Per cent	10.92	8.56	6.07	5.90	5.35	5.33	5.26
3. Other Uncultivated land excuding Fallow Land (A+B+C)	37.64	35.13	32.31	30.22	27.74	27.50	26.82
(A) Permanent Pasture and other Grazing Land	13.97	13.26	11.99	11.40	10.67	10.53	10.39
Per cent	4.25	4.03	3.65	3.47	3.24	3.20	3.16
(B) Land under Miscellaneous Tree Crops and Groves not included in Net Area Sown	4.46	4.37	3.58	3.82	3.44	3.45	3.31
Per cent	1.36	1.33	1.09	1.16	1.05	1.05	1.01
(C) Culturable Waste Land	19.21	17.50	16.74	15.00	13.63	13.52	13.12
Per cent	5.84	5.32	5.09	4.56	4.15	4.11	3.99

Contd...

Table 16.2–Contd...

	Classification	*1950-51*	*1960-61*	*1970-71*	*1980-81*	*1990-91*	*2000-01*	*2007-08*
4.	Fallow Lands (A+B)	22.82	19.33	24.55	23.37	25.07	25.90	25.15
	(A) Fallow Lands other than Current Fallows	11.18	8.73	9.72	9.66	10.29	10.56	10.34
	Per cent	3.40	2.66	2.96	2.94	3.13	3.21	3.14
	(B) Current Fallows	11.64	10.60	14.83	13.70	14.78	15.35	14.81
	Per cent	3.54	3.22	4.51	4.17	4.50	4.67	4.51
5.	Net Area Sown (6-7)	133.20	140.86	140.29	143.00	141.36	140.73	140.86
	Per cent	40.52	42.85	42.68	43.50	43.00	42.81	42.85
6.	Total Cropped Area (Gross Cropped Area)	152.77	165.79	172.63	185.74	185.34	188.29	195.83
7.	Area Sown more than once	19.57	24.93	34.63	42.74	43.98	47.55	54.97
8.	Cropping Intensity*	114.69	117.70	123.05	129.89	131.11	133.79	139.03
	Net Irrigated Area	24.66	31.10	38.72	48.02	55.13	56.92	62.29
	Gross Irrigated Area	27.98	38.20	49.78	63.20	76.19	78.42	87.26

Source: Directorate of Economics and Statistics (2010), Department of Agriculture and Cooperation.

Table 16.3: Types of Degraded Lands.

Sl.No.	*Type of Degradation*	*Area (million ha)*
1.	Water erosion	57.15
2.	Wind erosion	10.46
3.	Salt affected	6.32
4.	Acid affected	12.00
5.	Waterlogged	12.00
6.	Degraded Forest	24.90
7.	Others	8.60
	Total	131.43

Source: Ayyappan S. (2011) Paper presented in ICAR Governing Body Meeting.

technologies. Further, the probability of bridging existing yield gap is more from irrigated areas with the introduction of high level input intensive varieties and hybrids. In general, during recent past decade though the weather conditions have not been very favourable but increase in irrigated area as shown in Table 16.2, has supported against the erratic and uncertain monsoon conditions. The scope of increasing productivity level is more particularly in rice, wheat, maize, barley, gram, lentil and mustard as the gap in potential and actual yields is almost of 50 per cent. In two major staple crops namely; rice and wheat on an average about 21 and 30 qt/ha yield is being harvested against their potential of 47 and 50 qt/ha yield respectively.

Even today, a small group of scientists, innovative farmers and some extension functionaries and policy makers acknowledge the dangers of agro-chemicals and advocate the cause of organic materials in partial or total replacement of these chemicals. Whereas, a large and strong lobby of industrialists and group of administrators, policy makers, scientists, extension workers and farmers believe that the use of recommended level of agro-chemicals is indispensable for increased agriculture production.

Approaches for Sustainable Agriculture

In the narrated background, in future absolute dependence on the agro-chemical based farming for increasing the crop yields is not only dangerous due to various adverse effects to our environment, health hazards to human and animal population, but also the yield level is plateauing in major crops. Therefore, there is an urgent need to find and popularize an effective alternative to agro-chemicals, which can ensure increase in crop yields and at the same time sustain the ecosystem in natural condition. Since, the outcome of input intensive green revolution is before us which compels us to re-look the approaches being followed and find out the best alternative mechanism for sustainable agriculture development. *Sustainable agriculture is defined as successful management of resources for agriculture to satisfy changing human needs while maintaining or enhancing the quality of the environment and conserving natural resources.*

Therefore, for sustainable agricultural production, soil health has to be improved adopting various options feasible to area specific situations. The level of organic matter (humus) increases the availability of nutrients in a soil; hence, regular addition of manures will improve the physical conditions and biological activity and improved the plant-available nutrients. Number of options such as various composts, crop residues, green manuring, bio-fertilizers, sewage, sugar industry waste etc. are locally available to use them for improving soil organic content and soil nutritional level. Some of the approaches suggested for maintaining or enhancing soil health for sustainable agricultural development are:

- Change in land use policy to avoid diversion of prime agricultural lands to non agricultural uses.
- Checking of soil pollution on account of industrial wastes and indiscriminate use of agro-chemicals.
- Adoption of Location specific farming systems.
- Crop Diversification
- Production and promotion of fortified fertilizers to mitigate multi-nutritional deficiencies.
- Adoption of IPNM practices to sustain the agriculture production and maintain the soil health.
- Use of nutrients on the basis of soil test value.

Chapter 17

Agricultural Development and Livelihood Sustainability in the Upper Sindh Basin, Madhya Pradesh

Vishwambhar Prasad Sati*

Professor of Geography,

Government K.R.G. Post Graduate Autonomous College

Gwalior, Madhya Pradesh

ABSTRACT

This Chapter examines the agricultural development in the Upper Sindh Basin, a tributary of Chambal River in Madhya Pradesh and discusses the livelihood sustainability. This basin falls under the semi-arid climate and is one of the poorer regions of the state. Farming system is dominated by the traditionally cultivated rainfed subsistence cereal crops. The cereal crops such as wheat, barley, sorghum, pearl millet, pulses, oilseeds mostly groundnut and mustard, are grown during the two cropping seasons of *rabi* and *kharif*. The cultivation of fruits and off-season vegetables is considerably low. Livestock farming runs parallel with agriculture, while its contribution in generation of income is almost negligible. It is the main occupation and the principal means of livelihood of *Gujjars*, a land-less community that habitat nearby the jungle. Draught animals play an

* Author E-mail: sati.vp@gmail.com

important role in the agricultural operations and providing manure to agricultural fields as fertilizer. Two factors considerably affect the farming systems. The first is low availability of water and the second is infertile soil except along the course of the main rivers, the Sindh and its tributaries, where alluvial soil is found in the small patches. The scope of enhancing production and productivity of these crops is low, unless adequate irrigation facilities are provided. The geo-ecological conditions are highly favourable for the cultivation of groundnut, mustard and to certain extent of cultivation of citrus, guava, papaya, and mango and of off-season vegetables. The forest based non-timber products can largely be utilized as most of the area of the basin comes under the 'Madhav National Park'.

Keywords: *Farming system, Sustainable livelihood, Subsistence cereal crops, Livestock, Upper Sindh Basin, Madhya Pradesh.*

Introduction

Agriculture is the main occupation of the people of the 'Upper Sindh Basin' (USB). About 80.2 per cent working population is engaged either in agriculture activities or in its allied practices such as livestock rearing and milk production. But all the produces are consumed at the local level. On the other hand, no commercialization of crop produces has been taken place so far. The fact behind is in this the region is still lagged behind, in order to introduce modern innovation in the field of agriculture. The main factors influencing the crops and their production are (i) traditional method of agricultural farming, (ii) dominance of subsistence cereal farming, (iii) lack of improved seeds, fertilizers, and other innovative uses, (iv) infertility of soil (soil is mostly stony and laterite), and lack of irrigation facilities. Under these circumstances, the production and productivity of these crops are comparatively quite low, even unable to meet the two times meal of the rural population. The major crops grown in the region are wheat, barley, oilseeds, including mustard (*rai/sarson*), groundnut (*moongfally*), soyabeen, pulses such as pea, gram, red gram (*tuar*), green gram (*moong*), black gram (*urd*) and etc. Rice is grown in a small part of the basin, only in the areas, where ample water for irrigation is available. The cultivation of rice is practiced along the course of the Sindh River, wherever, alluvial soils are available. Along with these crops, sugarcane is also grown in some plain parts of the basin mostly along the course of the river. Except groundnut and soyabeen, all crops are grown for domestic consumptions and their commercial viability is jest negligible. In terms of production and productivity, groundnut and soyabeen crops are getting impressive position in the cropping pattern and the farmers are utilizing their cropland mostly for the cultivation of these crops. As a result of this, the basin is leading in terms of production and productivity of these crops in the central India.

The farming systems of the basin are characterized by the rainfed agriculture and the crops are grown mostly during the four months of monsoon season and also during the winters. The irrigation facilities are inadequate. The crops are consumed domestically. Meanwhile, horticultural crops including fruits, flowers, and vegetables along with groundnut and soyabeen, which have already attended tremendous position in the farming systems, can be grown on the course of the Sindh River with replacing millets. The purpose of the study is to examine the viability of cash crops in lieu of the subsistence cereal crops, which could manifest a way for sustainable livelihood. The main research questions were raised during the study is how far the cash crops other than cereal crops are economically viable and by which way it would by helpful to enhance livelihood sustainably.

Geographical Location of the Study Area

The USB is located in the northwest part of Madhya Pradesh state. The Sindh River has its origin at village Gopi Talai (Hatolion plateau, the west of Sironj) in Lateri taluk of Vidisha district at an elevation of 1750 feet (533.40 m). The Sindh is a tributary of the Yamuna River, flows mostly through MP state in districts of Vidisha, Guna, Shivpuri, Gwalior, Datia, and Bhind. It joins the Yamuna River near village Jagwanpur in Etawa district of Uttar Pradesh. The total length of the river from its origin to its confluence with Yamuna is 500 km. The command area is elongated and narrow with a length of about 95 miles (152 km) and average width of 23 miles (37 km). The total drainage area up to confluence with the Yamuna River is 6723 sq. miles (17413 sq. km).

The extension of the USB is limited to Narwar taluk of Shivpuri district and the total drainage area is 2139 sq. miles (5540 sq. km). It comprises the part of Vidisha, Guna, and Shivpuri districts and falls between 24° 2′ N to 25° 32′ N and 77° 22′ E to 77° 58′ E. It is characterized by the two distinct geographical regions namely; the Udaipur-Gwalior highland and the Great Malawa plateau (Ayyar *et al.*, 1971). Lateri and Sironj taluks of Vidisha district and Guna, Isagarh, Aron and Ashoknagar taluks of Guna district come under the Malawa plateau and all taluks of Shivpuri district such as Shivpuri, Pichhor, Khaniadhana, Kolaras, Badarwas, Karera, and Narwars come under the Udaipur-Gwalior highland. Shivpuri plateau covers the southern part of Shivpuri district (Bhattacharya 1971a). The important tributaries that join the USB are Barahi *nalla*, Kair *nalla* on the left bank and Aer *nala*, Iter *nalla*, and Chobch *nalla* on the right bank. Figure 17.1 shows the location map of the Upper Sindh Basin (USB).

The study area is economically backward and one of the poorer regions of the state. Climatic conditions are characterized by the semi arid climate. Rainfall occurs only during the four months of the summer by the monsoon of the Bay of Bengal. During the winter season, rainfall occasionally occurs due to the western disturbances, which is locally known as *'maotha'* (Sati and Gupta, 2004). Except, on the course of the river and some other small patches, where alluvial and black soils are found, the entire region is characterized by the infertile soil. Inadequate facilities of irrigation further affect low production and productivity of crops negatively (Sati, 2003). At the same time, the scope and economic viability of cash crops such as groundnut, soyabeen, mustard along with off-season vegetables and fruits are considerably high.

Major Crops Grown Under Traditional Agricultural Farming

The entire basin is characterized by the dominance of subsistence cereal crops, which possesses 71.6 per cent average area of the total cropped land. It varies from one taluk to another as Shivpuri taluk obtains lowest area under cereals (44.6 per cent) followed by Narwar (56.4 per cent), while Latery occupies 86.6 per cent, Sironj 82.7 per cent, and Aron 82.9 per cent. Badarwas has 74.9 per cent cropped land while Guna has 74.9 per cent, Pichhor and Khaniadhana have 62.8 per cent cropland each. Table 17.1 reveals taluk wise area under major crops (in per cent) in 2000.

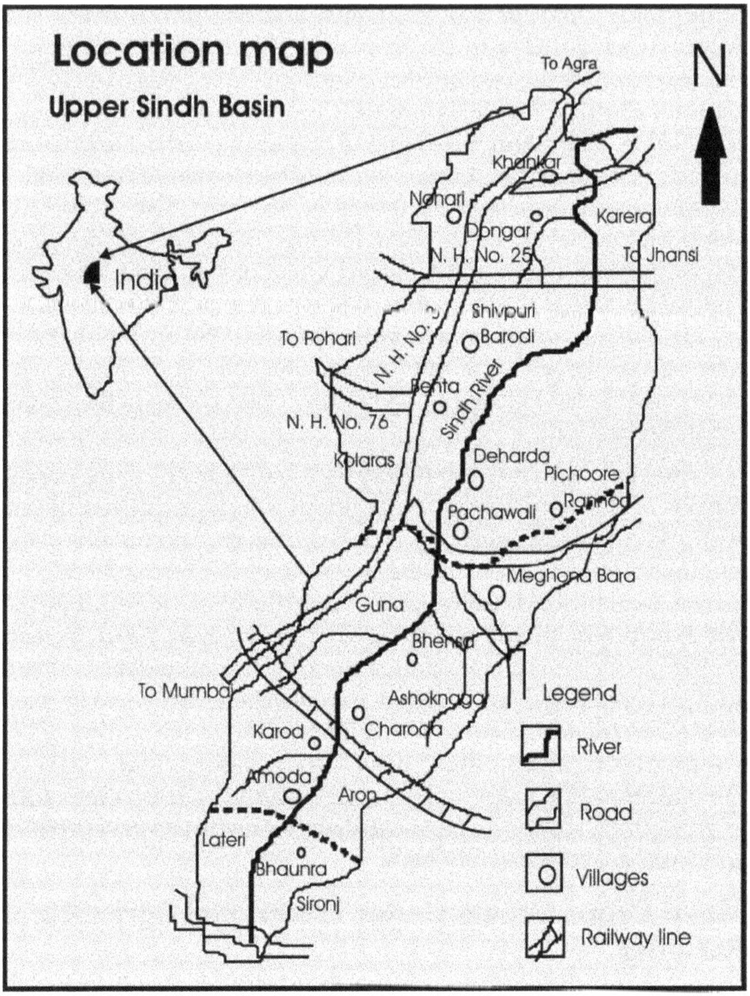

Figure 17.1: Location Map of the Upper Sindh Basin

Table 17.2 reveals net sown area, production, and productivity of the major crops grown in the basin. The highest net sown area is noticed under wheat crop (33.6 per cent) followed by gram (31.2 per cent) and then soyabeen (15.2 per cent). Similarly, in terms of production, wheat crop obtains first rank (559.8 thousand MT in 1998-99) followed by gram (255 thousand MT) and then soyabeen

Table 17.1: Taluk-wise Area Under Major Crops (in per cent) in 2000

Taluks	*Major Crops*					
	Wheat	*Jowar*	*Maize*	*Gram*	*Ground-nut*	*Soya-bean*
Shivpuri	3.8	0.2	0.1	2.3	1.4	6
Pichhor	3.8	0.08	1.3	1.2	4.1	1.7
Khaniadhana	4.0	0.4	2.1	2.6	2.7	2.6
Kolaras	2.6	1.7	0.7	16.3	0.1	9.2
Badarwas	3.8	1.2	0.4	10.2	0.1	4.4
Karera	6.0	–	0.7	1.7	7.5	0.5
Narwar	5.4	0.07	0.1	3.0	5.5	0.05
Guna	9.6	3.5	1.6	8.7	0.01	13.3
Ashoknagar	9.0	1.6	0.4	9.4	0.03	5.9
Isagarh	7.2	0.9	0.5	9.4	0.01	3.1
Aron	4.3	1.7	0.5	4.3	–	2.2
Lateri	5.2	0.4	0.3	4.7	0.0006	3.7
Sironj	7.0	0.6	0.3	10.8	0.06	5.7

Source: Statistical Diary of Shivpuri, Guna, and Vidisha Districts 2002.

Table 17.2: Net Sown Area, Production, and Productivity of Major Crops

Crops	*Per cent of Net Sown Area*	*Production (in Thousand MT)*		*Productivity (per ha yield in TMT)*		
		1995-96	*1998-99*	*1995-96*	*1998-99*	*Growth in Per cent*
Wheat	33.6	489.6	559.8	14.8	16.9	14.3
Rice	0.5	5.20	6.00	10	12	20
Jowar	5.5	100.3	40.0	20	8	–20
Maize	3.9	68.3	47.8	22	15.6	–31.8
Gram	31.2	183.6	255	5.9	8.2	60
Tuar (red gram)	0.3	1.70	1.10	3.3	3.2	–3.0
Soyabean	15.2	126.4	139.5	8.4	9.2	12.5
Groundnut	5.7	53.5	72.3	10	14.4	40
Linseed	0.7	3.2	2.7	4.2	2.8	–50
Mustard	1.4	2.10	1.6	2	1.1	–50

Source: Agricultural Department of Shivpuri, Guna, and Vidisha Districts, 2002.

(139.5 thousand MT). While, when we see increase in productivity, Gram (60 per cent) and groundnut (40 per cent) rank high. The trend of production of Jowar, Maize, Linseed, and Mustard is decreasing. It is quite interesting to note that the people are understood about the value of cash crops and they are replacing the farmlands of cereal crops. The description of major crops is as follows:

Wheat

Wheat obtains the highest rank in terms of production and area covered and it is one of the main cereal crops in the USB. About 33.6 per cent land is under the wheat crop while, it obtains about 71.7 per cent land under cereal crops. The premier production area of the wheat crop is Ashoknagar, Isagarh, Karera, Narwar, and Guna taluks. The highest area under wheat crop falls in Guna taluk (9.6 per cent) followed by Sironj (7 per cent) and Karera (6 per cent).

Jowar

Jowar is the second largest crop grown in the basin. It occupies 5.5 per cent land of the net sown area. It is mainly grown in Guna, Kolaras, Badarwas, Ashoknagar, and Isagarh taluks. Guna taluk occupies the highest percentage under Jowar crop (3.5 per cent) followed by Badarwas (1.7 per cent), and Ashokanagar (1.6 per cent).

Maize

Maize is grown only in 3.5 per cent of the total area. It is mostly grown in Pichhor, Khaniadhana, and Guna taluks with 1.3, 2.1, and 1.6 per cent area respectively.

Rice

Rice is grown in a very less part of the basin. It occupies only 0.5 per cent area and is grown only near the water bodies.

Gram

Gram is one of the most important crops grown in the USB. It occupies about 31.2 per cent area, out of the total gross sown area and 84.6 per cent area, out of the total area under pulses. It ranks second after wheat crop in terms of area. The highest area under gram falls in Kolaras taluk (16.3 per cent) seconded by Badarwas (10.2 per cent).

Soyabean

Soyabean is the main oilseed of the basin, which obtains 15.2 per cent land of the total gross sown area and 58 per cent land of the total area under oilseeds. Guna taluk placed first (13.3 per cent) followed by Kolaras (9.2 per cent) and Shivpuri (6.0 per cent) taluks in terms of area.

Groundnut

Groundnut is sown in 5.7 per cent cropped area. It has second place in area and production of oilseeds. This basin has the ideal conditions for growing groundnut. Karera and Narwar taluks lead in its area and production.

Mustard

Mustard is grown in only about 2.7 per cent cropped area. It occupies the third rank in area and production of oilseed. It is grown mainly in Shivpuri, Karera, and Narwar taluks.

Production and Productivity

In the USB, production and productivity of cereal crops is considerably low. It may be due to lack of adequate irrigation facilities, stony and barren land (soil infertility), and lack of modern innovation in the field of agriculture. The productivity of crops varies from one geographical location to other; as the areas, which fall in the upper-middle part of the basin, obtains high yield of crops. The reason is availability of black soil as this region is located in the Malwa plateau. The productivity can be noticed considerably high in the course of the Sindh River due to presence of alluvial soil and availability of water. Recently, the three phases project was proposed by the state government and out of them first phase (Mohani Sagar pick-up canal) is fully completed, which is resulted in providing adequate water for the cropped land. It is noticed that the areas, which are having adequate irrigation facilities, are able to grow high quality and quantity of rice and sugarcane. The second and third phases of the project have to be completed in the coming time. The second phase is devoted to generation of electricity and irrigation, while third phase is devoted to irrigation only. Successful completion of the project will definitely manifest to sustainable development of agricultural farming, which will reduce food scarcity and enhance livelihood options.

Strategies for Sustainable Livelihood

As much discussed in the paras cited above that the production and productivity of cereal crops, which are largely grown in the USB is considerably low. Similarly, the economic viability of these crops is also negligible. The major reasons of emerging this situation in the basin are (i) inadequate irrigation facilities and (ii) much talked low soil fertility. Irrigation facilities can be enhanced after completion of all three phases of the USB's project. It will also manifest a way for growing the cash crops including off-season vegetables and fruits on the traditional cropped land along with cultivation of subsistence cereal crops, according to the agro-ecological conditions. The economic viability of cash crops, which include soyabean, groundnut, mustard, off-season vegetables and among fruits; papaya, guava, mango, watermelon, and citrus fruits, is tremendously high,. Southern part of the basin, which borders with the Malwa plateau, has the privilege of having fertile black soil. Similarly on the course of the main river (Sindh), alluvial soil is abundant. Enhancing food security through adopting policies measures and their rigorous implementation will surely lead to the sustainable livelihood in this economically backward region. The following strategies can be adopted for the sustainable livelihood in the basin:

Development of Infrastructural Facilities

Proper irrigation facilities and easy access to market is the need of the hour. This is one of the big measurements, which can reduce the poverty and enhance the livelihood options. The successful completion of the project in different phases will generate hydroelectricity and water for irrigation facilities in the vast rainfed areas. Around 162, 100 ha cropped land is estimated to have canal irrigation facility after completion of second phase (Sati, 2002).

Promotion of Off-Season Vegetables

In the cropped land where water supply is available, particularly along the course of streams (major and minor) and canal side, off-season vegetables can be grown. The major vegetables, which can largely be grown in this basin, are potato, onion, cauliflower, radish, egg-plant, beans, lady finger, and cucumber. Extensive and intensive cultivation of vegetables, which are already grown in the basin, will lead to the sustainable livelihood. Table 17.3 exhibits that about 2047.2 *ha* land is under vegetable cultivation while production is 16028 tones. Three vegetable crops obtain highest area and production. Potato ranks first (662.6 *ha* land and 5960 tones production) followed by tomato and onion (23.7 per cent and 12.9 per cent land respectively).

Table 17.3: Area (ha) and Production (tone) of Vegetable Crops

Name of Vegetable	*Area in ha*	*Per cent of Total Area Under Vegetables*	*Production in tones*
Potato	662.6	32.3	5960
Tomato	487.8	23.7	3885
Onion	266.8	12.9	4496
Egg-plants	129	6.3	359
Cauliflower	119	5.8	1065
Lady finger	24	1.2	203
Arbi	24	1.2	60
Other	334	16.3	–
Total	**2047.2**	**100**	**16028**

Source: Horticultural Departments of Shivpuri, Guna, and Vidisha districts.

Cultivation of Fruits, Flowers, and Medicinal Plants

The main fruits grown in this basin are mango, guava, papaya, citrus, and orange. Although, the area under these fruits is almost negligible, as table 4 shows yet, the production and productivity is quite high in comparison of the cereals crops. The highest area under fruit crops is obtained by guava (588 *ha*) followed by mango (403 ha). Citrus ranks third with 192 *ha* area while papaya bears forth place (117 *ha*). Similarly in terms of production (1998-99), guava possesses first place with 10647 quintal production followed by mango (4777 quintal). Subsequently, citrus occupies third rank (3377 quintal) followed by papaya (2142 quintal). Recently, the need of cultivating flowers and medicinal plants was felt by the farmers and they started cultivating them. Government assistance to promote this practice as a subsidiary source of livelihood is required at every level.

Table 17.4: Area (ha) and Production (Quintal) of Fruit Crops

Name of Fruit	*Area in ha*	*Per cent of Total Area Under Fruits*	*Production in Quintal*		*Growth per cent*
			1997-98	*1998-99*	
Mango	403	24.0	1140	4777	319
Guava	588	35.0	6520	10647	63.3
Citrus	192	11.4	946	3377	256.9
Papaya	117	6.9	1735	2142	23.5
Orange	39	2.4	–	–	–
Others	340	20.3	750	3178	323.7
Total	**1679**	**100**	**11091**	**24121**	**117.5**

Source: Horticultural Departments of Shivpuri, Guna, and Vidisha districts.

Rearing of HYV Animals for Milk Production

Livestock farming is the subsidiary occupation in the basin. Every household has one or more livestock, particularly in the rural areas. The down stream part of the basin comprises Shivpuri, Karera, Kolaras, Narwar, and Pichhor taluks has the potential of milk production because there are extensive grassland and forestland are found, which is the main source of fodder. A large number of *Gujjar* community people are engaged in rearing of animals and producing milk. Attention should be given for rearing of HYV animals and infrastructural facilities should be provided at the village level.

Promotion of Forest Based Small-Scale Industry

Promotion of forest based small-scale village industry such as bidi industry, collection of gum, honey, and medicinal plants and making of wood craft and furniture will be helpful for enhancing livelihood options. This area represents one of the major *tendu leaves* (use to make bidi) producing areas of the state, while there is no bidi industry is established in the basin. India's biggest national parks 'Madhav National Park' and Kuno-Palpur are located in this basin. Therefore; small-scale forest based industry with optimum utilization of non-timber based forest products will generate the income and augment employment here.

Conclusions

The regional economy is dominated by agriculture practices as it engages 80.2 per cent of the total active workers and is the mainstay of the bulk of population (Bhattacharya, 1971b). The main occupation of the people is agriculture and seasonal labouring. Because unproductive nature of soil, low rainfall and lack of irrigation facilities, the production and per ha yield of the crops is considerably low in comparison to the other parts of the state. The subsistence cereal crops, which are the main base of livelihood, are not sufficient, even to meet the daily food requirement of the populace. To boost up the present systems of agriculture, the following measures can be adopted:

1. Traditional crops should be replaced by the cash crops, which are suitable in the agro-ecology of the basin. Such crops are soyabean, groundnut, mustard, off-season vegetables, fruits, medicinal plants, and flowers. As, it is experienced from the past decades that the production of these cash crops is getting impressive position in terms of productivity and land cover. *Rabi* crops have the equal importance that are largely grown in the southern part of the basin and include wheat and gram. A proportion of land should also be devoted to these crops.
2. Irrigation facilities can enhance the production and productivity of crops and in the coming fall, ample water supply will be provided for cropped land. Rational utilization of irrigation facilities, according to need base, will provide sustainable livelihood.
3. Agriculture is practiced by the traditional mode of cultivation. To raise crop intensity, production, and productivity; new innovation in the field of agriculture, keeping ecology of the region in view, should be introduced.
4. Composite manure should widely be used to keep fertility of soil maintained.
5. Optimum utilization of non-timber based forest products and introduction of HYV animals for milk production, according to area specificities, would be a milestone in the sustainable development of the basin.

Acknowledgements

This study was an outcome of the Pilot Study funded by 'the Indian Council of Social Science Research (ICSSR)'. I sincerely acknowledge ICSSR for providing financial assistance to commence this study.

References

Ayyar, N.P., Verma, K.N., Nigam, R.K. and Singh, K.Y., 1971. Malwa Plateau. In: *India: A Regional Geography*, (Ed.) R.L. Singh.

Bhattacharya, A.N. a&b, 1971. Udaipur: Gwalior region. In: *India: A Regional Geography*, (Ed.) R.L. Singh.

Sati, V.P., 2003. *The Upper Sindh Basin: A Study on Resources Analysis and Area Development*. Unpublished Pilot Study submitted to the Indian Council of Social Science Research. New Delhi.

Sati, V.P., 2002. *Environmental Impact Assessment and Socio-Economic Study of Madikheda Dam Project*. Unpublished Major Project Report submitted to UGC, New Delhi.

Sati, V.P. and Gupta, L.D., 2004. *Shivpuri Jile Ka Bhogolik and Arthik Vishleshan*. Pratik Prakashan, Shivpuri.

Chapter 18

How Climate Change Affects India

Sunil M. Gawande[1] and and Vijay Singh Rathor[2]*

[1]*Research Scholar, R.T.M. Nagpur University, Nagpur, Mh.*

[2]*Assistant Professor, Department of Geography,*
Government P.G. College, Morena, M.P.

ABSTRACT

Climate change is one of the most important global environmental challenges, with implications for food production, water supply, health, energy, etc. Addressing climate change requires a good scientific understanding as well as coordinated action at national and global level.This paper addresses these challenges. Historically, the responsibility for greenhouse gas emissions' increase lies largely with the industrialized world, though the developing countries are likely to be the source of an increasing proportion of future emissions. The projected climate change under various scenarios is likely to have implications on food production, water supply, coastal settlements, forest ecosystems, health, energy security, etc. The adaptive capacity of communities likely to be impacted by climate change is low in developing countries. The efforts made by the UNFCCC and the Kyoto Protocol provisions are clearly inadequate to address the climate change challenge.

This article is focus on the India's interest to ensure that the world moves towards a low carbon future. Many studies have underscored the nation's vulnerability to climate change.

* Corresponding author: E-mail: sunilgawande99@gmail.com

With changes in key climate variables, namely temperature, precipitation and humidity, crucial sectors like agriculture and rural development are likely to be affected in a major way. Impacts are already being seen in unprecedented heat waves, cyclones, floods, salinisation of the coastline and effects on agriculture, fisheries and health. India is home to a third of the worlds poor, and climate change will hit this section of society the hardest. Set to be the most populous nation in the world by 2045, the economic, social and ecological price of climate change will be massive.

Introduction

Precisely at a time when India is confronted with development imperatives, we will also be severely impacted by climate change. Like other developing countries, several sections of the Indian populace will not be able to buffer themselves from impacts[2,8] of global warming. With close economic ties to natural resources and climate-sensitive sectors such as agriculture, water and forestry, India may face a major threat, and require serious adaptive capacity to combat climate change. As a developing country, India can little afford the risks and economic backlashes that industrialized nations can. With 27.5 per cent of the population still below the poverty line, reducing vulnerability to the impacts of climate change is essential.

It is in India's interest to ensure that the world moves towards a low carbon future. Many studies have underscored the nation's vulnerability to climate change. With changes in key climate variables, namely temperature, precipitation and humidity, crucial sectors like agriculture and rural development are likely to be affected in a major way. Impacts are already being seen in unprecedented heat waves, cyclones, floods, salinisation of the coastline and effects on agriculture, fisheries and health. India is home to a third of the worlds poor, and climate change will hit this section of society the hardest. Set to be the most populous nation in the world by 2045, the economic, social and ecological price of climate change will be massive.

Major Facts on Climate Change in India

The future impacts of climate change, identified by the Government of India's National Communications (NATCOM) in 2004 include:

- Decreased snow cover, affecting snow-fed and glacial systems such as the Ganges and Bramhaputra. 70 per cent of the summer flow of the Ganges comes from melt water
- Erratic monsoon with serious effects on rain-fed agriculture, peninsular rivers, water and power supply
- Drop in wheat production by 4-5 million tones, with even a 1°C rise in temperature
- Rising sea levels causing displacement along one of the most densely populated coastlines in the world, threatened freshwater sources and mangrove ecosystems
- Increased frequency and intensity of floods. Increased vulnerability of people in coastal, arid and semi-arid zones of the country
- Studies indicate that over 50 per cent of India's forests are likely to experience shift in forest types, adversely impacting associated biodiversity, regional climate dynamics as well as livelihoods based on forest products.

India stands to lose on too many counts to allow a 'climate-politics-as-usual' scenario. Therefore, positive engagement with global climate negotiations at the next UNFCCC meeting in December 2009 is crucial.

India's Accelerating Emissions

Although not an emitter historically, India currently has one of the fastest growing economies in the world. With a government target of 8 per cent GDP to achieve developmental priorities[16], a share of one sixth of the global population, and changing consumption patterns, India's emissions are set to increase dramatically.

Growing at an almost breakneck pace, and guzzling coal, gas and oil in large quantities[4], we are today, the fourth largest emitter of greenhouse gases worldwide. Although our per-capita emissions are among the lowest in the world, our growth rates imply that the past is no predictor of the future[8]. The most recent IPCC report suggests that India will experience the greatest increase in energy and greenhouse gas emissions in the world if it sustains a high annual economic growth rate. The International energy Agency predicts that India will become the third largest emitter of greenhouse gases by as early as 2015.

India imports large quantities of fossil fuels to meet its energy needs, and the burning of fossil fuels alone accounts for 83 per cent of India's carbon di-oxide emissions. Nearly 70 per cent of our electricity supply comes from coal.

Although India has maintained its clear economic and social development imperatives, the government recognizes that climate change is an serious problem, and that business as usual is no longer the way forward.

Climate Change Perception

Unsustainable Consumption Patterns

Unsustainable consumption patterns of the rich industrialized nations are responsible for the threat of climate change. Only 25 per cent of the global population lives in these countries, but they emit more than 70 per cent of the total global CO_2 emissions and consume 75 to 80 per cent of many of the other resources of the world (Parikh *et al.,* 1991). In per capita terms, the disparities are also large: an Indian citizen emits less than 0.25 tonnes of carbon per year whereas a citizen of the USA, for example, emits more than 5.5 tonnes.Parikh *et al.* (1991) emphasise the need for an equitable and efficient solution to climate change and suggest that efficiency can be obtained through a system of tradable emission quotas and equity through equal allocation of global environmental space to all human beings. These findings were well received by developing countries and are echoed in the UNFCCC, which recognised the rights of developing countries to economic development and also the "common but differentiated responsibilities" of different countries.

Methane Emissions and Subsistence Emissions

Another theme of Indian analysts has been the lack of reliability of GHG emission estimates, particularly of methane. According to initial estimates, large emissions of methane from paddy fields were ascribed to developing countries. However, the empirical basis of these estimates was questioned; subsequently experimental measurements by Indian researchers showed these doubts to be well-founded (Mitra, 1992,1996). Moreover, Agarwal and Narain (1991) argue that emissions by poor who live on t he margin of subsistence should be considered a basic human right and should not be counted when ascribing responsibilities for emission reduction.

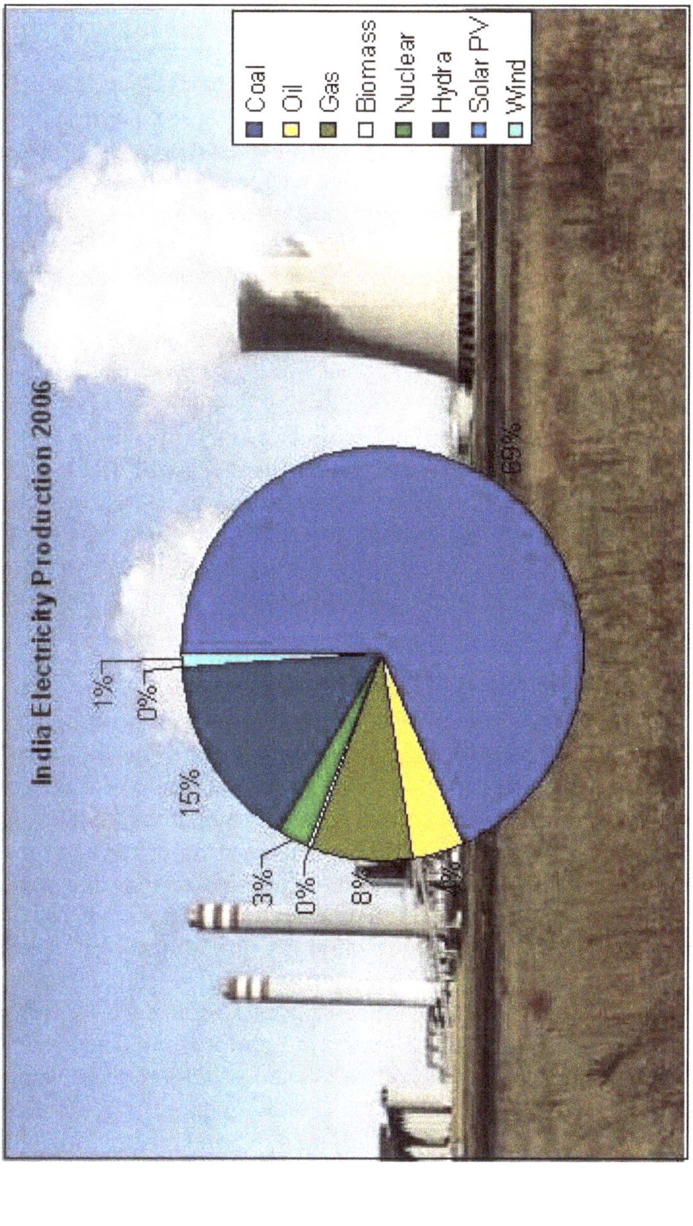

Figure 18.1: India Electricity Production 2006

Why Should India be Concerned about Climate Change?

Indians should be concerned about climate change since this phenomenon might have substantial adverse impacts on them. Not all possible consequences of climate change are yet fully understood, but the three main 'categories' of impacts are those on agriculture, sea level rise leading to submergence of coastal areas, as well as increased frequency of extreme events. Each of these pose serious threats to India. However, these are long term issues. The overriding immediate concern for India should be the fast pace at which negotiations are taking place on the climate front. India's main energy resource is coal. With the threat of climate change, India is called upon to change its energy strategy based on coal, its most abundant resource, and to use other energy sources (e.g. oil, gas, renewable and nuclear energy) instead, which may turn out to be expensive. Thus, an immediate issue is to come up with a better negotiation strategy such that we have more freedom to decide which type of energy we use, how we generate power, how to reduce methane emissions by agricultural practices or forestry and so on. Negotiations are important for us as a means to reduce or postpone future vulnerability by getting the developed countries to reduce their emissions.

Risk of Lower Agricultural Production

The FCCC objective states that GHG concentrations should be stabilized at levels where food production is not threatened (UN, 1992). Thus, by examining the impact on agriculture of different climate change scenarios, one can get an idea of what is tolerable. Rosenzweig and Parry (1994) have estimated significant adverse impact on the agriculture of many developing countries. In a more detailed study of India, Kumar and Parikh (2001a and 2001b) examined the impact of climate change on agricultural crop yields, GDP and welfare. Considering a range of equilibrium climate change scenarios which project a temperature rise of 2.5°C to 4.9°C for India, Kumar and Parikh (2001a) estimated that: (a) without considering the carbon dioxide fertilization effects yield losses for rice and wheat vary between 32 and 40 per cent, and 41 and 52 per cent, respectively; (b) GDP would drop by between 1.8 to 3.4 per cent. Their study also showed that even with carbon fertilization effects, losses would be in the same direction but somewhat smaller. Using an alternative methodology Kumar and Parikh (2001a) showed that even with farm-level adaptations, the impacts of climate change on Indian agriculture would remain significant. They estimated that with a temperature change of +2°C and an accompanying precipitation change of +7 per cent, farm level total net-revenue would fall by 9 per cent, whereas with a temperature increase of +3.5°C and precipitation change of +15 per cent, the fall in farm level total net-revenue would be nearly 25 per cent. For a developing country, these are very large changes which can cause much human misery. From India's point of view, a 2°C increase would be clearly intolerable. Other developing countries may be even more vulnerable (possibly Bangladesh or Small Island States).

Risk of Sea Level Rise

Large-scale emigration from coastal zones is expected due to submergence of coast-lines after sea levels have risen. This will create large numbers of environmental refugees especially from low-lying delta regions in poor countries. Furthermore, intrusion of sea-water in the groundwater and changes in temperature can reduce agricultural and fishing incomes. Countries dependent on coastal fishery and agriculture, which most often include developing countries, are likely to be adversely affected. If a one-meter sea level rise were to take place today, it would displace 7 million persons in India (ADB,1995). In the future many more may be displaced. 35 per cent of the land in Bangladesh would be submerged by a one-meter rise. The estimates for costs to build walls along the zones vulnerable to sea level rise for the USA is $107 billion in 1989 prices (Yohe, 1990). That may be a small share of the GDP of developed countries, but such measures, even scaling for their coast lines, for say, Bangladesh, could require a very large share of its GDP. Who shall pay Bangladesh or India for such a wall? Given that these countries are unlikely to be able to pay for protective measures, tens of millions of people will be displaced in Bangladesh and many of them could spill over into India.

Risk of Extreme Events

Increased occurrence of extreme events due to climate change will also affect the poor most. In the cyclone in Andhra Pradesh in India in 1996, more than 1,000 people died and there was huge property loss. Cyclones of similar intensity in advanced countries like the U.S. may not lead to any deaths and much hardship, due to stable and durable housing and other infrastructure and extended safety net available to the people in distress.

The Costs of Changing Energy Strategy

If India has to reduce its carbon emissions, it would mean a major reorientation of her energy strategy, especially if that warranted a shift from its current coal-based to a oil and gas based energy system. Murthy, Panda and Parikh (2000) estimated costs associated with a low GHG energy strategy in terms of foregone income and welfare of the poor. They examined the consequences of alternative CO_2 emission reduction strategies on economic development and, in particular, the implications for the poor by empirically implementing an economy-wide model across India over a 35-year time horizon. A multisectoral, inter-temporal model is used for this purpose. The model has specific technological alternatives and endogenous income distribution with dynamic behaviour; it covers the whole economy in an integrated top-down-bottom-up model. The results show that CO_2 emission reduction imposes costs in terms of lower GDP and higher poverty. A 30 per cent CO_2 reduction over a period of 30 years using annual emissions reduction targets leads to a fall in GDP of 4 per cent and raises the number of poor by 17.5 per cent in the 30th year (that is, if 2000 were taken as the baseline, these changes would occur by 2030). Cumulative emission reduction targets are, however, preferable to annual reduction targets and an optimum strategy where the country is free to decide when to reduce emissions as long as over 30 years the same amounts of cumulative emissions are reduced. It reduces the hardship of emission reductions with a cumulative reduction target of 30 per cent (*i.e.* not specifying annual targets), the fall in GDP is 1.4 per cent and increase in the number of poor is 6 per cent in the 30th year both of which are less than in the case of annual reduction. The scenarios involving compensation for the loss in welfare are not very encouraging, as they require large capital inflows. The required minimum compensation to maintain welfare levels is $278 billion for annual reduction and $87 billion for cumulative reduction respectively. The payments are large, and are concentrated in the early years of the period concerned. A more realistic strategy would call for compensation payments which are spread out over the years even though it may require larger total inflow over the years.

Conclusion

India has committed to actively engage in multilateral negotiations in the UNFCCC, in a 'positive and forward-looking manner'[15]. The government recognizes that 'global warming will affect us seriously' but maintains that the 'most important adaptation measure to climate change is development itself'[8]. This has ensured that India's position at the UNFCC has stubbornly remained 'common but differentiated responsibility'. Under the UNFCCC agreement itself, India is not subject to any binding emission reduction targets until the year 2012.

In spite of this guarded stand, India has 'declared' that even as it pursues its social and development objectives, it will not allow its per capita emissions to exceed those of developed countries. The 11th 5-year plan does make headway in reducing energy intensity per unit of GHG by 20 per cent while boosting cleaner and renewable energy.

In June 2008, the Prime minister released the much awaited National Action Plan on Climate Change (NAPCC). The NAPCC outlines a strategy by which India will adapt to climate change, while maintaining a high growth rate, protecting poor and vulnerable sections of society and achieving national growth objectives[11]. It focuses on eight areas intended to deliver maximal benefits to development and climate change (mitigation and adaptation). However, detailed action plans for each mission, and any clear targets are missing from the report.

Although the action plan may be a missed opportunity for leadership on climate change, the good news is that change is coming[8]. Realizing that the market is changing, and not to be left behind in the global race, Indian businesses are beginning to take on climate change as a business issue.

What we need now is for the Government of India to capitalize on India's position as a developing giant, take the lead and engage with Governments of the world and the private sector for a low-carbon future.

Contrasted with these, scenarios involving tradable emission quotas give India an incentive to be carbon efficient. It becomes a net seller for the first 25 years, say up to 2020, if the quota is in the range of 1 ton per person as per 1995 population. Because of the reduction in carbon intensity due to the incentives provided by tradable quota, less is emitted in later years when India becomes a net buyer. The results suggest that for India, and other developing countries, the window of opportunity to sell carbon quotas is only in the next two decades or so. Thus, precious time is lost during prolonged negotiations.

References

http://www.indiaclimateportal.org/how–climate–change–affects–india (Access on October 05, 2011).

http://www.oecd.org/dataoecd/22/16/1934784.pdf (Access on October 05, 2011).

http://www.ias.ac.in/currsci/feb102006/314.pdf(Access on October 05, 2011).

Chapter 19

Natural Resources Utilization and Management for Sustainable Development: A Case Study of Himalayan Watershed

*B.L Teli**

Department of Geography, H.N.B. Garhwal University Campus-Pauri, Garhwal, U.K.

ABSTRACT

The essence of natural resource growth lies in a happy blend of the realization of the physical outer limits to the exploitation of environmental resources and the inner limits of human needs and aspirations. If we mismanage the resources, we have to face several environmental problems at various stages. Thus, the resources use status, reflect our development and improvement goes to every hand. One has to use natural resources to control the environmental hazards and damages like pollution, soil erosion, desertification, deforestation, floods and other ecological imbalances. The development is not only based on per capita income, consumption and standard of living but also on human development for moral and ethical

* E-mail: blteli33@gmail.com

foundations, to provide balanced economy and resource sustainability for healthy environment to grow. Since the resources of the earth are limited. There is no disputing the fact that the conservation of natural resources is basic to sustainable development. Therefore, we must develop a holistic concept of development by ensuring both environmental protection and sustainable availability of resources for maintaining the growth in production over population growth. It requires united effort on the part of professionals to evolve appropriate technologies and strategies, which are environment-friendly, socially equitable and economically viable. Development which is not equitable can not be sustainable in the long term. So, we have to come up with the strategies of mutually reinforcing actions at individual, local, regional, national and global level for protecting the system sustainability and maintaining the sustainable development.

The population explosion along with technological development, urbanization and materialistic nature of human being during the last century have upset the man-resource balance. Developing countries like India should not be parasite on the developed countries in reference to the resource and environmental management. The control of environmental pollution and/or degradation is the main issue. Due to adoption of modern means of science and technology, the man is exploiting/modifying the resources according to his own need and requirements. With the results the ecological system has mostly been jeopardized. The existence of life has been put into danger. Both the natural and cultural hazards have forced the large scale resources degradation and ultimately ecological imbalances. The natural resource degradation included floods, droughts, desertification, water erosion and atmospheric pollution while cultural hazards include over population, slums, poverty, deforestation and pollution; the degradation is increasing the trends in desertification and waste land and deteriorating the agricultural products even in such areas where once the flourishing lands were found.

The global resources are facing a number of problems due to increasing biotic pressure, deforestation, over exploitation of land resources and pollution. The natural resources over exploitation can be observed in the form of soil erosion decreasing productivity of agricultural land, climatic changes, lack of forests and forest products, decreasing capacity of pastures etc. Not only these, but the pollution is at the all time high, has not only polluted the earth but affected the upper strata as well by depleting the ozone (O_3) layer. It may have serious implications on the human health, his activities and plants. It may also reduce crop yields significantly.

The developing countries face a near crisis situation, economic as well as environmental. The Himalayas are no exception of it. In this region the natural balance between environment forming and conserving forces on the one hand and degrading and depleting on the other, has been disturbed. It is the function of soil-water forest conservation, population and pollution control measures to restore this balance, even if the face of ever increasing pressure on the resources to provide food, fodder, fuel, industrial raw material etc.

The resources of the earth are limited. The population explosion and technological development has upset the man-resource balance. Of recent years there has been a growing awareness of, and interest in, the problem presented by environmental degradation. Much of this interest is stimulated by issues involved in planning a sustainable mode of development which must be environmentally friendly, economically viable and socially equitable. more recently, there has been a spectacular expansion of interest in sustained economic growth and problem of making judicious use of natural resources such as soil, water, forest etc. progressively deteriorating ecosystem as a result of biotic and abiotic stresses is posing great challenge to the life support system and consequently the sustainable development at different spatial scales. There is no disputing fact that the conservation of natural resources is basic to sustainable development.

Developing countries like India should not be parasite on the developed countries in reference to the resource and environmental management. The essence of environmental growth lies in a happy blend of the realization of the physical outer limits to the exploitation of environmental resources and the inner limits of human needs and aspirations. If we mismanage the environment, we have to face several environmental problems at various stages. Thus, the environmental status reflects our development and improvement and goes to every hand. One has to control the environmental hazards and damages like pollution, soil erosion, desertification, deforestation, floods and other ecological imbalances. The development is not only based on per capita income, consumption and standard of living but also on human development for moral and ethical foundations, to provide balanced economy and environment.

The control of environmental pollution and or degradation is the main issue. Due to adoption of modern means of science and technology, the man is modifying the environment according to his own need and requirements. With the results the ecological system has mostly been jeopardized. The existence of life has been put into danger. Both the natural and cultural hazards have forced the large scale environmental degradation and ultimately ecological imbalances. The natural hazards included floods, droughts, desertification, water erosion and atmospheric pollution while cultural hazards include over population, slums, poverty, deforestation and pollution; the environmental degradation is increasing the trends in desertification and waste land and deteriorating the agricultural products even in such are as where once the flourishing lands were found.

Global Scenario

The global environment is facing a number of problems due to increasing biotic pressure, deforestation, over exploitation of land resources and pollution. The environmental degradation can be observed in the form of soil erosion decreasing productivity of agricultural land, climatic changes, lack of forests and forest products, decreasing capacity of pastures etc. Not only these but the pollution is at the all time high, has not only polluted the earth but affected the upper strata as well by depleting the ozone(O_3) layer. It may have serious implications on the human health, his activities and plants. It may also reduce crop yields significantly.

The developing countries face a near crisis situation, economic as well as environmental. The Himalayas are no exception of it. In this region the natural balance between environment forming and conserving forces on the one hand and degrading and depleting on the other, has been disturbed. It is the function of soil-water forest conservation, population and pollution control measures to restore this balance, even if the face of ever increasing pressure on the environment to provide food, fodder, fuel, industrial raw material etc.

Every year more then 1000 animals' species and 20,000 flowering plant species are in danger of existence and more than one third of the known bio-species have already been lost. A number of planning's and technological inputs used for human welfare and becoming an environmental problems due to the population explosion, illiteracy, poverty, economic and social inequality, materialistic and consumption oriented life system and the industrialization has polluted the total environment. Hence, the useful life saving system has been narrow down, polluted and poisoned. Earlier the effects were limited to wildlife, plants and natural landscape only but now a days the environmental degradation, Physical and Chemical pollution and resource deterioration is seen clearly every where.

The human economic activities have affected the 20,000 crore hectare land bought under agricultural uses. It has lost 2 crore sq. km area, which is more than the present agricultural area. The human activities such as-construction, mining, soil erosion, desertification and salinity take away as much as 50,000 to 70,000 sq. km. agricultural land every year. Only the erosion take away 250 crore tones of soil every year. We have already lost 5 lac sq. km area of cultivated and pasture lands under desertification. Every year the urbanization, in the developing world, take away 3,000 sq. km area. Carbon-di-oxide (CO_2) has increased by 36 lac tones while the 24 lac tones oxygen (O_2) has been lost from our environment. The temperature is increasing and the problem of green house effect is felt. Ozone (O_3) layer has lost 7 per cent over the Antarctic an environment and it may further loose in the coming year.

The third most deteriorating affect of the industrialization is, 'Acid Rains' which has imbalanced the total environmental system. About 40 per cent of the total permanent water supplying drainages are converted into waste lands. Every year, more than 60 lakh hac. Forests are cut down for various purposes. So, we are loosing 100-200 crore tones oxygen (O_2) every year. According to the 'Red Data Book', 400 Birds, 138 Amphibians, 305 mammals and 193 fish species and sub-species are in danger. Similarly 25,000 plant species are also in danger. According to another report approximately 20,000 plant species and 900 animals' species of chordate group are in danger.

The ecological imbalance and environmental degradation is perceptible in the form of excessive silt load in river channels and torrents during rains, eroded and denuded hills as a consequence to over grazing, low percentage area under forest cover with much less density then optimum, gradual disappearance of humans from forests, lack of natural regeneration, drying of perennial springs and early drying of ephemeral springs, reduction in miscellaneous and broad-leaved species with ecological retrogression, an increase land slides, rock falls and debris, wild life is being fast exterminated and both man and mountain are becoming poorer. Indiscriminate road construction has done a lot of damage of forests. The road passes through virgin where "U" bends are created to gain altitude as the roads ascends, extension of cultivation on steep slopes, natural calamity in monoculture forestry cover free grazing of animals, collection of fuel, fodder etc. for domestic needs and cooking food, etc. help in degradation. The civil and soyam and panchayat forests have completely disappeared. The erosion of cultivated fields, community lands, wasteland and land disturbed due to various activities in the name of development add very high amount of silt to flow with water. The mining and quarrying activities in the outer hill ranges are glaring examples of deforestation and damage of natural resources.

So, the problem is more serious in a mountainous region where physical configuration-nature of geological formation, altitude, relief, slope, drainage characteristic and biotic pressure has led to the environmental degradation only10 per cent of the world population live is mountain areas. But another 40 per cent live in adjacent plains, so the life and livelihood of half the world, directly, depends on the way in which the watershed ecosystems are managed.

Environment of Study Area

The western Nayar is the tributary of main Nayar River which pour water to the river Ganga from the left bank. The basin covers 745 sq. km area. The river originates from the Dudhatoli range which forms the border of Garhwal, Chamoli and Almora districts. It originates at the heigh of 3015 m and confluence the main channel near Satpuli (Bhattkoti-villages) at the height of 560m, 92.5 per cent area ranges between the heights of 800m and 2400m. The region is bi –sected by the main channel in northern and southern parts. It can further be divided into 20 sub-watersheds which have as many as 3343 segments (2583 I order) under different orders. The total length of the streams is calculated to be 2.96 km./sq. km. On an average slope is 27.7.

The watershed receives 1302mm average rainfall per annual in as many as 76 rainy days. The co-efficient of variation is calculated to 46.8 per cent. The region has sub-tropical to cool climate. It changes with the change in altitude, direction of rides and their aspects, intensity of forest cover etc.

The western Nayar basin has a total population of 117311 having a growth rate of 14.31 per cent sex ratio is 1223 females/ 1000 male, literacy rate is 38.26 per cent the workers percentage is 38.44 more than 80 per cent are engaged in primary section only, as many as 24131 house holds with average size of 4.86 per cent person/ household stay in 599 settlements; Average size of settlement is 195 person/settlement. The density of population is high 157 persons per sq. km and S.C. population is 15.2 per cent.

Forested area covers 42.01 per cent cultivable land 37.4 per cent blank land is 20.6 per cent area. Further 7.65 lands are not available for cultivation, 9.3 per cent pastures and grazing land, cultivable waste land, is 11.6 per cent. Most of the agriculture is limited to the valleys and slopes are terraced for the same. Horticulture has low extension in the region.

The watershed has as many as 1272391 fields covering 25898 hectare area calculated to be 2.03sq.m/field. There are as many as 388547 land owners having on an average 33 fields fragmented at least four to 6 sites. The number of cultivators is 36982 and cultivated area is 25898 ha., hence, each cultivator has 0.70 ha. Due to the biotic pressure on the land every possible piece of the land owned by a farmer has been brought under plough. The intensity of cropping is 152.4 per cent while the intensity of irrigation is 196 per cent on 3.2 per cent of N.S.A. Mandua, wheat and rice are the main crops. The productivity is low and can feed only 55 per cent of food requirements (147.104 kg/per/annuam). The technological inputs are very-very limited in the region.

There are 12, 2,696 farm level animals in the region which are calculated to be 84304 animals units. Cow, Bullocks, Buffaloes and Goats are the main live stocks which area of local quality and produce a very low amount of milk etc. The density of live stock is 168 units/

sq. km man livestock ratio is 0.72 cattle units per person. Regarding the fodder supply the region can feed 75 per cent animals and remaining animals require 50,000 tones of fodder every year which has degraded the reserved forests of region very badly. Regarding the fuel wood it can feed only 1/3 of the requirements and the resources are so exploited that they can not provide the timber.

Thus, extension of cultivation on steep slopes, free grazing, natural calamities, collection of fuel and fodder for domestic needs and fire etc. help in promoting the environmental degradation. Apart from these, construction works also help in salutation and degradation. Indiscriminate road construction

Construction has done a lot of damage to the forests and fertile land. Every km of road produces 550 Quebec meter debris by land sliding and rock fall etc. killing vegetation, fertile soils and chocking mountain streams as well as down streams to create floods in the plains and siltation in reservoirs. Hence the mountains are becoming poorer. The run-off measured at Jwalpa Devi is 2.85m^3/sec.

The region has four genetic types of soils namely red loams, brown forest soil, pod sol soils and meadow soil. The nutrients in the soil are found medium *i.e.*, Nitrogen is 2.94 per cent. Phosphorus 2.71 per cent, potassium 2.75 per cent on an average. There are three main soil forming rocks-biotic schist, granite gneiss and phyllites. Based on general soil characteristics, soil profile, texture and slope, the store Index were calculated to be 61.2 per cent.[3] The region can be divided into three main pedo-ecological zones:

Table 19.1: Pedo-Ecological Zone of W. Nayar Basin

Sl.No.	*Pedo-Ecological Zone*	*Altitude (m)*	*Climate-Vegetal Zone*	*Soil Type*
1.	Lower Valley Zone	Up to 900	Humid Tropical	Alluvial soils
2.	Middle Zone	900-1800	Warm Temperate	Brown forest soils
3.	Upper W. Nayar Zone	< 1800	Cold Temperature to cold	Brown deciduous and gray coniferous forest soils.

Physical Configuration and Natural Resources

Altitude varies from 565m. At the confluence point to 3015 m at the highest peak of the region. The upper part known as Dudatoli area has altitude above 1600 meters. The central part of the basin has low altitude while peripheral parts have high altitude and channels originating from them poor there water to the main channel from north and south. The upper W. Nayar basin above the height of 2000m. Covers 22.2 per cent areas of the basin while the central part ranging between 1200 and 2000m. Embraces 62.8 per cent area and lower basin cover only 15 per cent area which is below the height of 1200m. It is evident from Figure 19.1 that there are a number of intermittent ridges in the basin and give rise to the streams. There are as many 20 sub-watersheds and 3343 segments in the region under different orders. The bifurcation ratio is calculated to more than 4.0.

Table 19.2: Area Under Different Latitudinal Zones

Sl.No.	*Altitude(m)*	*Area (sq. km.)*	*Per cent to total area*
1.	Below 1200	113	15.0
2.	1200-1600	245	33.1
3.	1600-2000	222	29.7
4.	2000-2400	124	16.7
5.	Above	24	5.5

The western Nayar catchments lying entirely within the lesser Himalayas show a high altitudinal difference and a number of erosional and depositional landforms of glacial glaciological fluvial origin. The topography of the basins is highly rugged-horned peaks, serrated crests of high ridges, cirques, hanging valleys, torrential Rapides, gigantic escarpments and gorgeous topography. "U" and "V" shaped valleys are found. The flat bottom valleys development is associated with the development of terraces and fans, though they are not regular and extensive

The upper slopes have generally horn and cirque on the both sides of the ridges. Along the main central water divide where the lithologies are weaker like shale, phyllites etc. the mass movement in the form of creep is common while in the Dudatoli area formed of metamorphic rocks, the physical weathering is the dominant and

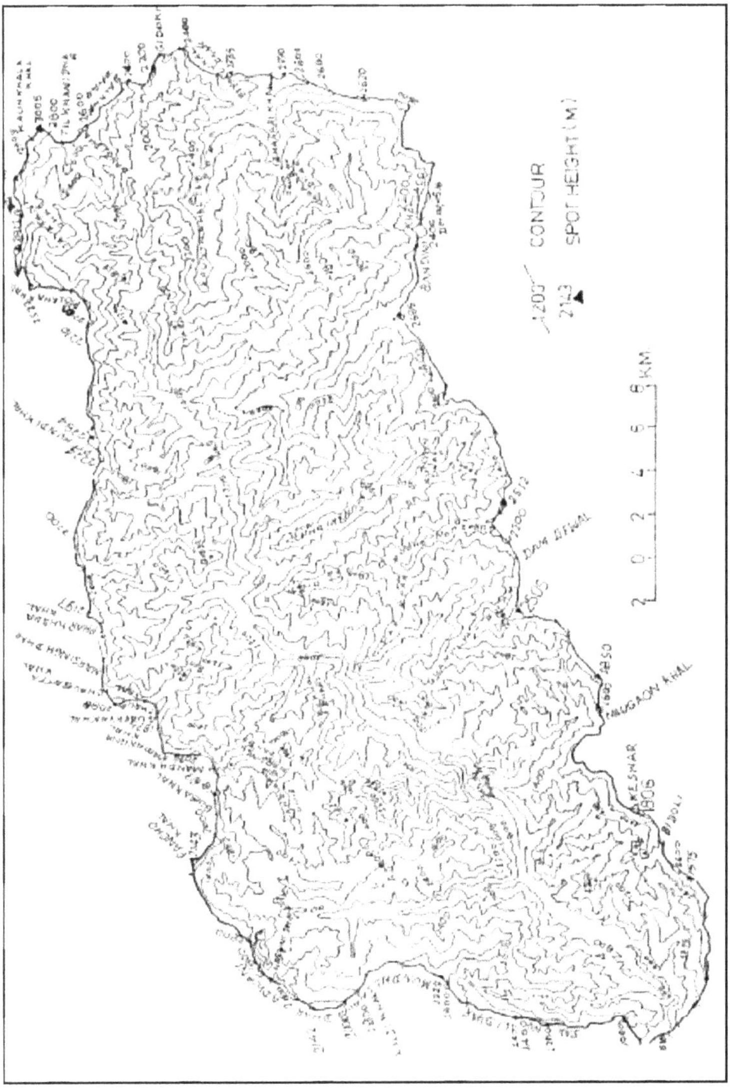

Figure 19.1: W. Nayar Basin (Contour Map)

deposited along the middle parts of slope. Moderate slopes are mainly confined to flat top ridges near Buakhal, Damdeval, Rauthia, upper parts of Kantli gad, and along Ekeshwar- Naugaonkhal water drive, represent mainly erosional features. The nature of slope is considered as index soil erosion. Conceptually with the increase in slope the intensity of soil erosion also increases. The data reveals that 32 per cent land surface is facing the problem of severe erosion while, 11.5 per cent is moderately eroded in the watershed.

Table 19.3: Slope Magnitude, Soil Erosion and Run-off

Sl.No.	*Slope in Degree*	*Area per cent*	*Nature of Soil Erosion*	*Intensity of Run-off*
1.	Below 20	11.50	Slight	Slight
2.	20-30	56.50	Moderate	Moderate
3.	30-40	32.00	Severe	High
4.	Above 40	00.50	Very intence	High

Superimposition of streams and channels over the landscape is another phenomenon parallel to soil erosion, as the controlling factors of the two are almost the same. The length of channels increases with the increase in head ward and downward erosion. It can thus be summarized that the length of stream and channels in an area can be considered as an index to soil erosion. The precise relationship has not been established, but there has been an agreement in considering drainage texture as an index to soil erodibility.[4] The drainage texture is expressed as the length of streams per unit area, while its reciprocal is the distance between the two adjacent channels, half the distance of later represents the length of over-land flow from divide to the channels[5], which again is important for the soil erodibility estimation. The present study is also based on the similar lines but the length of streams per sq. km. area has been called here as drainage density.[6]

The calculated values of drainage density in the region varies from 0.75 to 5 km/sq. km. The critical value of drainage length per sq. km. to warrant soil erosion is 0.96 km and over. On an average, 1 sq. km of drainage area is sufficient to produce and maintain 1.4 km of channel.[7] A lower value of drainage length then the referred value (0.96km./sq.km) is therefore, safe: it approaches the threshold limit

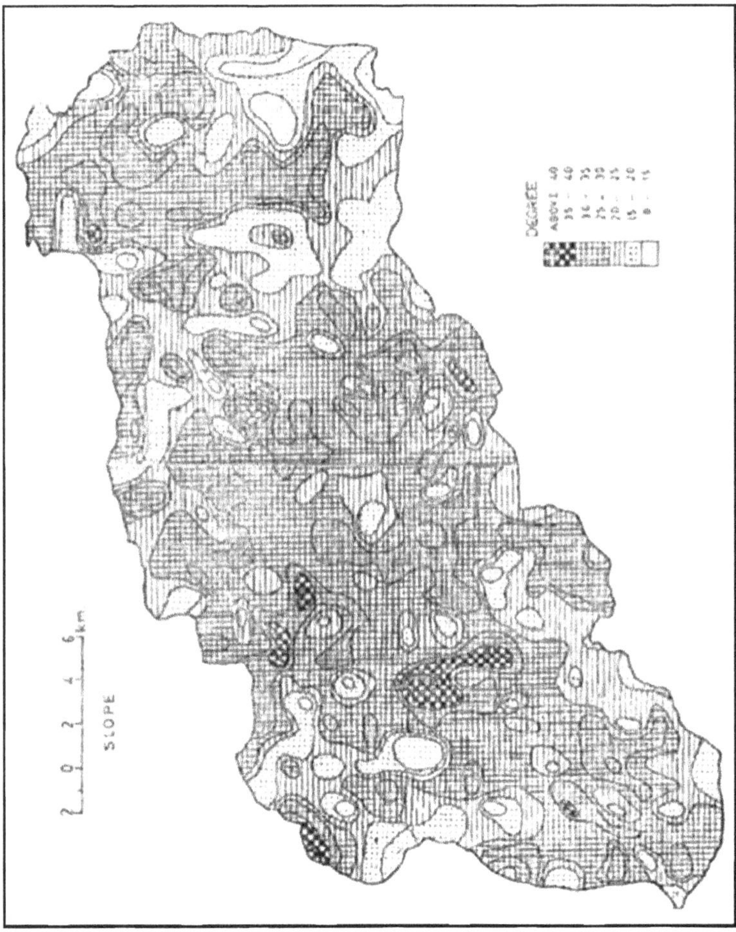

Figure 19.2: W. Nayar Basin (Slope)

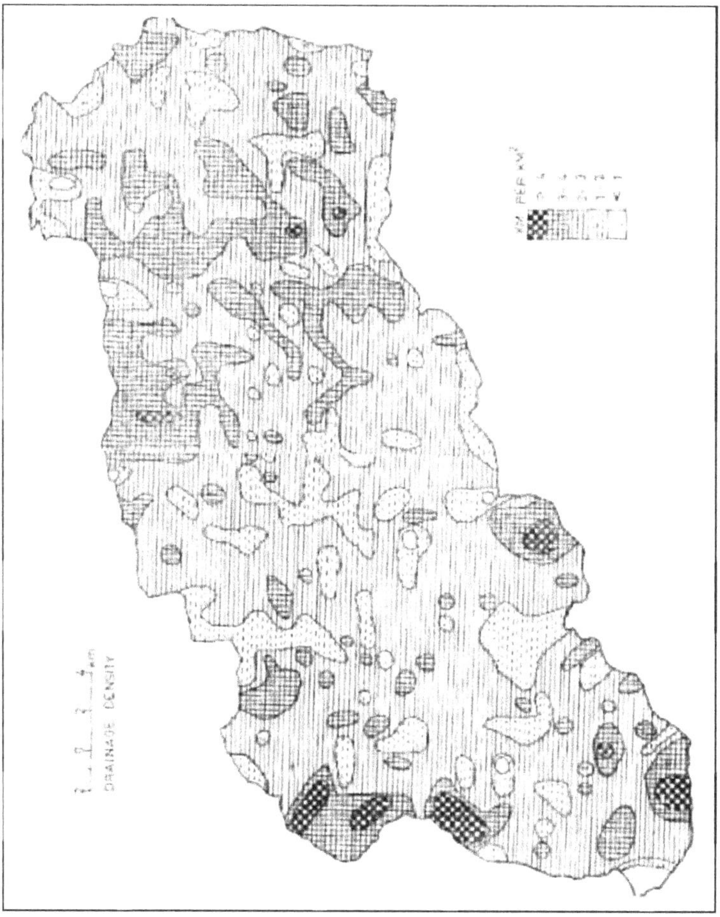

Figure 19.3: W. Nayar Basin (Drainage Density)

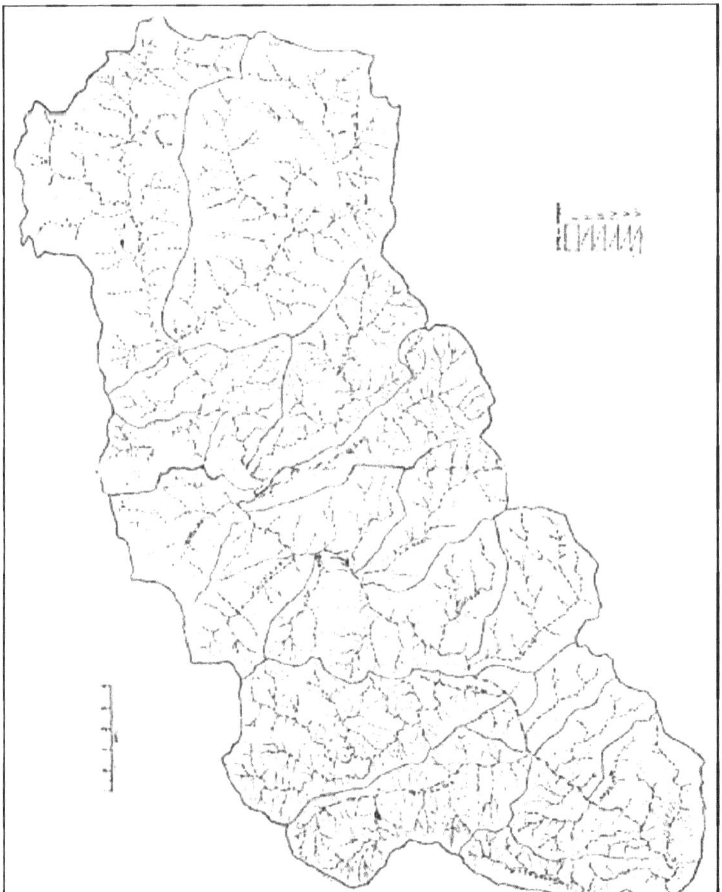

Figure 19.4: W. Nayar Basin (Drainage)

of erosion; values higher than this will automatically make for greater soil erosion.

None of the parts of the region have the drainage density below the critical value. Therefore the region is susceptive to produce more soil erosion. The region has on an average 2.96km. Length of segments per sq. km which is evident from the above discussion that none of the part of region is free from the erosion. In other words, the region produces more erosion than the natural erosion.

Low soil erosion is observed in the Sigad and its adjacent area and at the down reaches of the Khar gad and Amotha gad along the main channel (W. Nayar). Another region of this category is formed around Musagali-Chaplori-Tall-Champeshwar-Paboli settlements in the south of the main stream. The upper zone of the basin, the ridge area between the Dhaijyuli and the Syoli gad, also form the region of less erosion. Apart from this, a few parts on spurs and in valleys having coarse texture also have less erosion. The measured area under this category is 131sq. km.

Almost all parts of region experience moderate erosion. It reveals that wherever the drainage texture is moderate, length of slope is more and the erosion intensity is between 2 and 3 kms/sq. km. It covers almost 46.61 of the total area.

The zone of high erosion are mainly confined to the upper-middle zone of the basin covering lower part of Dhaijyuli gad, Baligad basin and upper reaches of the Chhoya gad, Bidoli gad, Pedul gad, Ghat gad Khar gad basins. The upper reaches of Sigad and Amotha gad also from the zones of high soil erosion.

Table 19.4: Area under Different Erosion Intensities

Sl.No.	*Index of Soil Erosion*	*Area sq. km*	*Per cent to Total*	*Nature of Erosion*
1.	Below-1	10.5	1.41	Very low
2.	1-2	130.2	17.60	Low
3.	2-3	347.8	46.61	Moderate
4.	4-4	224.0	30.04	High
5.	Above-4	32.0	4.29	Very high

The areas of intense erosion (32sq. km.) are localized in small patches conceding very fine drainage texture. Such patches are found around the Adwani in the east, Chamma Danda in the south, and in upper middle zone of the main basin.

Drainage density is the true index of soil erosion and environmental degradation of a region. Even if, there is a steep slope, it may not produce more soil erosion if it is covered with natural vegetation. Therefore, apart from the geological nature of rocks, natural vegetation controls, reforms and develops the length of stream in a unit area; and due to increase in the length of segments or channels, the erosion intensity along the channel length increases. Therefore, the first order segments in the region have more density and covers area which is liable to produce more run-off and soil erosion (Table 19.5).

Table 19.5: Stream Order, Length and Intensity of Erosion

Stream Order	*Area per cent*	*Length of Streams km/sq. km*	*Nature of Erosion*
1.	60.31	5.75	Very High
2.	29.27	2.80	Medium
3.	6.35	2.43	Medium
4.	3.70	1.02	Low
5.	0.37	2.17	Medium
6.	100.00	2.96	Medium to High

Apart from slope, land use plays an important role in the soil erosion and its conservation. Here, the erosivity of land under various land used have been calculated.

Table 19.6: Land Use and Soil Erosion (Area in sq. km)

Intensity of Erosion	*Forests*	*Cultivation*	*Barren Land*	*Total*	*Per cent to total*
Slight	178.7	3.4	2.1	185.2	24.8
Moderate	93.3	261.8	21.2	376.3	50.5
Severe	40.2	12.1	128.4	180.7	24.3
Intense	0.4	0.4	2.0	2.8	0.4
Total	278.7	278.7	153.7	745.0	100.0
Per cent	42.01	37.38	20.61		

Soil Degradation

Soil is the most important ecological factor. It is a result of the action and reciprocal influences of parent rocks, climate, topography, plants, animals and age of the land. It can be represented as:

$$S = (g.e.b.)\,\Delta t$$

where,

S: Soil

g: Geology

e: Environment

b: Biological influents

t: Time.

The problem of soil erosion, land wasting and soil duplication is as old in fact as old as the first man cut the first tree to practice arable farming. The soil erosion has created a number of other related problems, - Siltation of reservoirs, affected water supply and transport, flood, droughts and water logging etc.

The erosion of soil is mainly created by the natural factor and man. The natural agents are water, wind, glacier and gravity of the earth and the human misbehavior which promotes to soil erosion, deforestation, misuse of pastures and unscientific agriculture, etc. The soil erosion is the creeping death of earth, man, culture and his environment.

The present rate of soil erosion is over 2500 million tones per year *i.e.* our half a tone of soil is lost by every man, woman and child on the planet. India contributes 18.5 per cent of the total soil loss at the global level while has only 2.4 per cent of the land area of the world. Throughout the ages, since the earth was formed, surface materials have been removed by powerful erosive agents (running, water, wind, ice). The normal or geological erosion takes place when ground is covered with vegetation and amount of soil removed is equal to the amount of soil formation. But when man interface with the course of nature the rate of soil erosion is greatly accelerated. The generalized and the estimated soil loss under forest is 90 kg per hac. Per year, under hill paddy it is doubled and if the same plot is continuously cultivated for a number of years the rate of soil loss may be as high as 32 times per year.

Running water is the greatest erosive agent as it is present everywhere in this hilly region. If a heavy shower falls directly on the bare earth surface it loosens the soil particles and runs off the surface in a series of small 'gadheras' (rival let) into carrying soil with it. Rain drops beating power is high when the ground surface is bare otherwise intercepted by trees, plants, grass etc. In the course of time, the gadheras deepened and widened into 'gad' (channels) or deep gullies (N.W. and S.E. parts). In extreme cases, it has given rise to highly dissected badlands. Such process is more common on cultivated steep slopes.

The rain water on steep hill sides also causes mass movement of the soil due to the lubricating action of water within the soil. Soil creep- the gradual downward movement in which large part of the hillsides slum down, are more common on bare steep slopes. The land sides and sinking of land near Simkhet, Pokharikhet and Narsingh Danda area (Siku-Chaupriyum), etc. are some of the shining example of it. The banks of the stream get eroded every year by the torrents and in wide valleys the streams change their course gradually (at the down stream of the main channel) every year.

Usually on landscape of gentle slopes in the region sheet erosion occurs. The top fertile layer of soil is ripped off in thin films every year by the running water from over the entire surface of such lands. Every time when it rains, the running water becomes muddy because of fine practical of soil which from the thin top layer of soil are taken away down to the lower reaches of channels. For some years landscape looks unchanged but due to the soil nutrients loss, yields of crop lower down slowly. When the soil becomes stony the gradual process of sheet erosion becomes perceptible gradually, the running water forms rills and then gullies which promotes further erosion over the land.

Hydraulic pressure exported by moisture penetrating into the soil during heavy rains and, thus, being unable to percolate further down because of impermeable rock strata, promote slip erosion when a huge landmass slides down bodily along the steep escarps. In the region such processes are more common, particularly along the roadsides. It is estimated that if heavy rains double the water flow scouring capacity is increased four times, carrying capacity by 32 times and size of the particles 65 times.[8]

Table 19.7: Lithology and Structures for Different Types of Movements

Types of Movements	*Nature of Movement*	*Slope*	*Favorable Lithology*	*Favourable Structure*	*Location*	*Remark*
Land slide	Rapid	Moderate to sleep	Weaker states, phyllites, schist etc. with beds of harder lithology	Planar structure large scale facet thrusts lineaments, lineaments, cleavages	Simkhet	Clay content becoming plastic on moisture
Rock slide	Rapid	Very steep	Hard rocks, quartzite, limestone, etc.	Bedding, joints deformation planes	Mandakhal	Excess rains
Block slide	Rapid	Moderate to steep	Hard rocks, quartzite etc.	Bedding joints, etc. (shattered and crushed)	Nautha, Tarpalisain	Physical weathering, excess rain.
Rock fall	Sudden steep	Moderate to	Hard rocks, quartzite limestone	Bedding joints etc. (shattered and crushed)	Kantligad Mandakhal	Detaching of rocks-mechanical
Creep	Slow	Moderate to steep	Weaker, surfacial, deposit terracettes	Planes, formed among the movements of water seepage.	Ekeshwar Nagdeo area Chau-rikhal etc.	Clay plastic and viscous of wetting frost action also.
Slump	Slow	Moderate to gentle	Superficial deposits terraces	Plane at contact of bed rock and the superficial deposit seepage.	Jwalpa Devi area	Excess water making the material viscous.
Debris flow	Slow	Moderate to slow	Debris, slope creep fans, talus etc.	Plane at contact of bed rock and rock debris/ scree	Nagdeo Amotha gad Sarnauligad	Water seepage-rains
Sheet slide	Slow	Moderate to steep	Weathered rocks on the top parts	Heavier precipitation, seepage along the contact of fed.	Buakhal, Kaljikhal, Thandhar, top flate ridges	2-3m thick weathering and the water move-ment

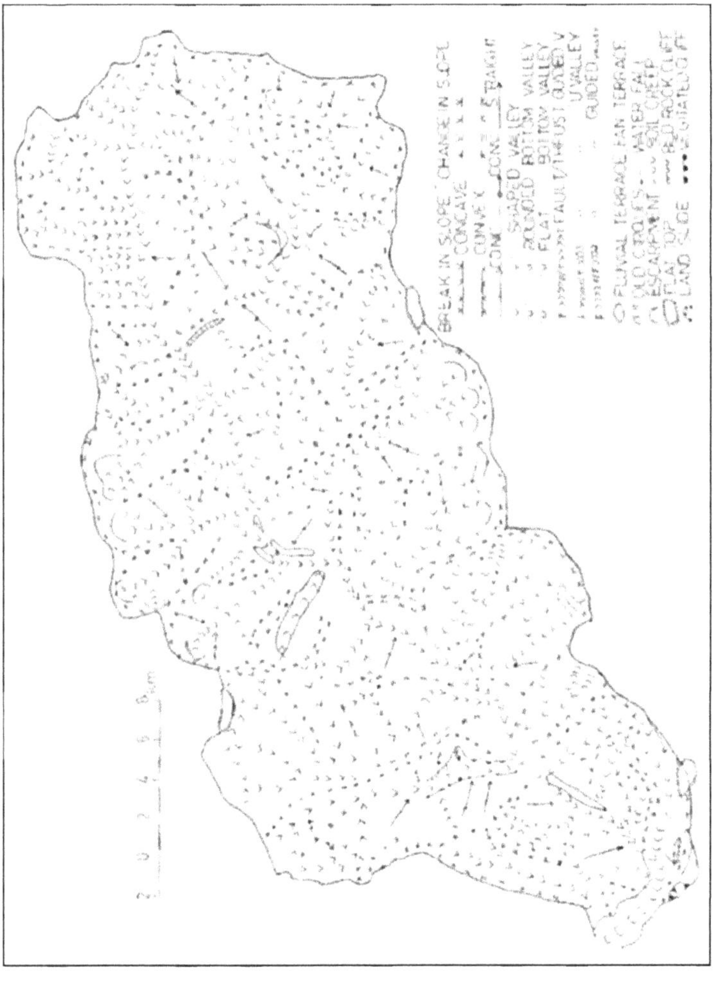

Figure 19.6: W. Nayar Basin (Geomorphology)

Biotic Pressure and Natural Resources Degradation

The biotic pressure has considerable influence on environmental degradation which becomes problematic with human and animal interference. The indiscriminate felling of trees, land disturbed for developmental activities, over grazing of pastures, small, uneconomic and fragmented fields, etc. promotes to denude the uppermost and protective cover of the surface. Not only that, but the footpaths used by the man and his cattle also develop into rills and gullies, causing 'Trail' erosion. This type of erosion therefore, is very common in most part of this hilly region.

Man and cattles, particularly his pastoral flocks and hards grazing round the year disturb the environmental balance. Forests provide basic soil nutrients but the local people have the habit to pilfer away trees of selected species for their use without bothering about their regeneration. With the result the forest remain for name sake in records but have deteriorated beyond repair. The gray poles of plant stems, now a days, look like ghost forests.

Even mildly sloping pastures which are overgrazed exhibit the sign of sheet erosion. Grasslands are more susceptible to landslides than forest lands on account of shallow root system. They are unable to bind and hold the soil as efficiently as trees and shrubs. The village pastures are overgrazed and the cattle remove every green blade of grass as fast as it emerges, will starve out. The vicious cycles continues until the land becomes a dreary waste fit to be abandoned. Annual fires followed by intensive rains are the principal causes of erosion and environmental degradation.

Free-grazing cattle scramble over steep hill sides in search of food, cutting the soil with their hooves, trampling down tree seedlings and leaving grass free tracks down which monsoon rain washes the loosened soil. The poverty of hill pastures is such that each animal unit needs at least 2-3 hectare grazing area to support the good health without causing environmental deterioration. While, the present pressure is 12 animal units per hectare of pasture and grazing land.

The pressure of increasing population is neutralizing all effects to raise the standard of living and of nutrition, while loss of fertility is nullifying the value of all improvements made. The population of our country as of March 31, 1997, was 94,98,9400. The annual growth rate is 1.93 per cent based in this growth rate India's population in the year 2050 will be 261,62,69,169. With the increase in population the demand for residences will continue to increase and thus the land available for agriculture progressively is reduced and the consequent result will be huge shortage of food stuff. In the villages the demand for fire wood will rise to such level that one wonders if any tree will be left in coming centuries. The spare potable water is not ever adequate for the existing population. From where will we get potable water for coming generation? It require measures for controlling population and bringing it to zero population growth rate (ZPG) in the year 10 to 20 years like we two our's two. And in the coming 50 years it should be taken down to one baby per family to prosper the ecological balance of this earth.

High literacy rate requires more advanced activities than the agriculture. Though agriculture too requires the technological know how but as such, the hill agriculture is an arduous and unprofitable proposition. Due to the lack of industrial and service Opportunities, most of the people are forced to leave their natives and approach to the plains to earn their bread and butter; mostly male members migrate, leaving families in the hills- women, children and old men. If the families were to accompany their men folk the hills would be almost empty. This region provides maximum human resources for Indian defense service. Most of the migrated people do not go back to their native places as they try to settle in foothills and other Indian cities to provide batter way of their coming generations. However, the migrated men in service send their earning home by money order. Purchasing grains, clothes, contingent and other family requirements. Due to men migration sex ratio is high in villages. Ladies look after the family and agriculture is well. The females of this region are hardworking and enthusiastic. This unidirectional migration of the people is creating one of the major obstacles in the development of this region. The money order and import of grains from Indian plains has permitted the residents to continue with their traditional farming practices. Had this region been as an isolated state, such an escape from environmental compulsions would not have been possible.

Land Use and Land Degradation

The watershed has 745 sq. km area out of which forest course 42.3 per cent whereas net sown area is 25.54 per cent (19000.6 hectares), agricultural fallow land is calculated to be 2.24 per cent and cultivable waste land is as high as 11.6. The availability of per capita net sown area is as low as 0.16 ha. Since, there is no scope for extending the cultivable land beyond the present limit; one has to depend on the present status of land for his livelihood. It is likely to diminish with the increase in population.

The pressure of population on net sown area is as high as 6.16 persons per hectare while the carrying capacity is only just half of it. The agriculture density or cultivated area per cultivator is 0.51 ha or 195 persons per sq. km. As it is clear, that the agriculture is the main stay of the rural masses in the basin. But on account of small, terraced, scattered fragmented fields and traditional mode of agricultural production with low output from cultivated land, it has failed to support the overwhelming majority of regional population. The high concentration of primary works in agriculture makes the sector weaker while low concentration provide the side-income to an agricultural family and make the sector better off. Therefore, is a need to diversify a fraction of primary workers to secondary and tertiary occupations?

By and large all agricultural land is terraced having slopes greater then 20° - 30° and is effected by moderate to severe erosion. According to the World Bank reports 15° slopes get arable with sound rotational practices, if bench terracing is done, for 25° slopes terraces are recommended and for 35° slopes orchard terraces or platforms are recommended. Land beyond 30° slopes thus should not be under cultivation. And on this basis at least 75 per cent of land of 30-40° slopes class needs change in land use.

Soil and water resources degradation and looses in the hilly terrain on different physic-cultural environment may be understood by the following examples.

It shows that bare fallow land produces maximum run-off while bare fallow and ploughed produces maximum soil loss. And grass cover and natural grasses produce minimum run-off and soil loss. (Table 19.8) Not only this but farming system and cropping pattern also help in soil and water loss in hills.

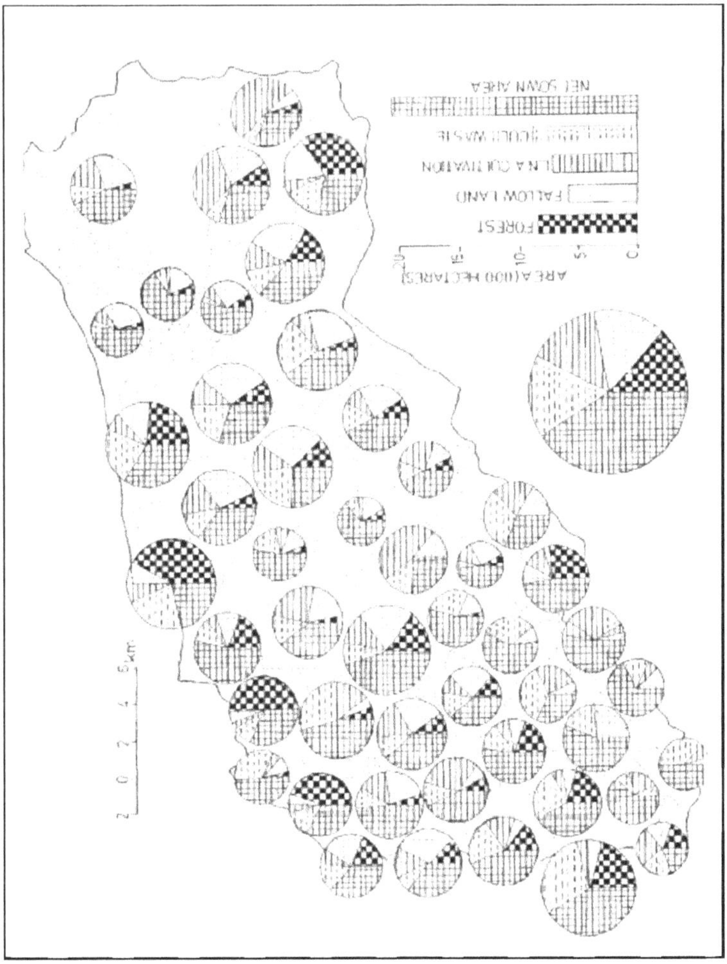

Figure 19.7: W. Nayar Basin (Land Use)

Table 19.8: Land Use-Run-off and Soil Loss on 9 per cent Slope

Treatment	*Monsoon Rains (June-Oct.) (Rainfall–1250 mm.)*		*Winter rains (Nov.-Mar) (Rainfall–173 mm.)*	
	Run-off as % of Rainfall	*Soil Loss t/hac.*	*Run*off as % of Rainfall*	*Soil Loss t/hac.*
Grass cover (*Cynodon dactylon*)	27.1	2.1	11.4	0.2
Bare fallow	71.1	42.4	58.7	3.6
Bare and ploughed	59.6	156	45.6	5.8
Natural grasses	21.2	1.0	4.1	0.0

Table 19.9: Farming System and Cropping Pattern– Soil and Water Loss on 8 per cent Slope (Rainfall 1240mm)

Treatment	*Water Loss (mm)*	*Water Loss as % of Rainfall*	*Soil Loss T/hac/year*
Up and down cultivation (Maize-Wheat)	670	54.1	28.5
Contour cultivation (Maize-Cowpea-Wheat)	511	41.2	19.3
Cowpea-Wheat	405	32.7	28.3
Cultivation fallow	445	35.9	44.0
Giant star grass	31	2.5	1.3

The nature of slope not affect the amount of soil and water loss only but it also weaken the soil nutrients and ultimately to its productivity. As it is evident from the Table 19.10.

Over 50 per cent of agriculture in the region is not justified on sound and husbandry principles. Agriculture in this area cannot give optimum returns and presently agricultural production is not at all commensurate with the framers inputs which are supplemented by government inputs in the shape of subsided for terracing or land shaping, supply of fertilizers, etc. Inspire of this, the subsistence farmer is UN willing to change and is locked into an economic dilemma from which he perhaps is unable to escape.

Table 19.10: Slope Length, Inclination and Cultivation Method and their Effect on Nutrient Loss (Kg./ha)

Treatment	Organic Matter	Total N	P_2O_5	K_2O	CaO	MgO
1. Slope length in feet						
60	41.0	3.8	1.8	7.3	20.0	4.3
120	34.0	3.2	2.2	7.6	20.6	8.4
180	33.4	2.7	3.0	12.5	25.7	12.9
2. Slope per cent						
0.5	86.8	5.8	10.7	42.8	53.8	41.4
1.5	92.8	6.5	11.1	52.9	59.2	78.5
3	173.9	10.8	23.5	117.8	203.2	211.8
3. Method of cultivation						
Contour rowed	25.4	1.41	0.058	0.984	37.8	2.4
Normally cultivated	158.5	8.81	0.366	6.15	236.2	15.0

Deforestation

There is pressing local demand for food, firewood, fodder, floods etc. and commercial requirement for herbs and other forest products, furniture and other use also deteriorate the natural forest. Apart from these, the forest area is encroached for cultivation, road construction etc. for the name of development.[9]

The upper valley and mid slopes of the watershed, where the density of settlements is high, forests are more degraded. Not only has this but the villagers claimed their right even in the reserved forest to fulfill their daily requirements and encroachment of others is considered to be the unavoidable for them. Even the govt. rules and settlement reports too could not change since 1865 [10]. The problem of fodder demand for existing stall fed cattle has already resulted in destruction of existing fodder forest. It takes the best part of half a day of a woman keep the cattle fed. The competitive scramble to collect fodder leaves, which may be fetched from a distance of several kilometers, hill women are hacking of large branches while previously used to remove only leaf bearing twigs. The fields are deprived of litter and forest floor eroded down to the tree roots. The free – grazing cattle has already, deteriorated the local forest-panchayat, civil and soyam forest, beyond repair. The cattle trample the over grazel pastures and seeding while scrambling over steep hillside in search of food. Hence farm- fodder forest linkages are imbalanced.

Apart from these, the demand of floats from specific trees and fuel requirements has also degraded the quality and quantity of forests. All the way every where, now a day, we see the chirpine forest, where, a few years back that area was covered by Oak trees-life, blood of hills. Oak has low generation power and capacity, the chir forest has come-up, which is considered to be not good for hill eco system. Forests are brunt annually to get fresh grasses, thereby accelerating the process of ecological degradation. The result of this denudation is that villagers are unable to obtain the needed fodder, firewood and timber. There is a vicious circle in which forests are getting depleted day by day and population pressure is increasing. Consequently the Oak forests have depleted, soil has degraded, springs have dried up and there is at places shortage of even drinking water.

Resources Availability and Utilization Status

The region has 84306 animal units. The total demand of fodder resources is 1981144 metric tones/year. The supply of fodder resources come from the cultivable land, civil and soyam forest and blank lands, which are available for collection as well as grazing. Considering the maximum level of supply is 126890 M.T/year only. It is only the 64 per cent of the total demand. Therefore, the remaining requirement of fodder is fulfilled from the reserved forest zones which ultimately help in environmental degradation.

There are as many as 24131 families in the region. The total demand of fuel wood is calculated to be 86872 MT/year at the rate of 3.6 MT/family/annum. The region can supply only 30622 MT fuel wood annually from forest and cultivated land. The deficit of fuel wood is as high as 65 per cent of the demand which is collected from the nearby reserved forests.

Similarly, the cultivated land can feed only 50-60 per cent demand of food stuffs. The remaining food resources are imported from the nearby plains.

Table 19.11: Production of Basic Resources in W. Nayar Basin

Source	Area (hac)	Yields T/hac./Year			Total Production T/Year		
		Fuel	Timber	Fodder	Fuel	Timber	Fodder
Forests	8326	1.0	0.5	2.5	8326	4163	20815
Grass	44592	0.5	–	1.75	22296	–	78035
Cultivated Land	14020	–	–	2.0	–	–	28040
Waste Total	66038	–	–	2.0	30622	4163	126890

Table 19.12: Demand and Supply of Basic Resources in W. Nayar Basin

Material	Population	Per Unit Use	Demand	Supply	Surplus/Deficit
Grass and tree fodder	84304 Au.	2.35 MT/year	198114 MT/year	126890.0	–71224
Food	117311 Persons	195.275 Kg/year	22899.1 MT/year	17266.9	–5632.2
Fuel wood	24131 Families	3.6 MT/year	86872 MT/year	30622	–56250
Timber	24131 Families	0.1 m^3/year	2413.1 m^3/year	41630.0	–2413.1

Table 19.13: Status of Land Resources and their Use (Area in sq. km)

Intensity of Erosion	Forest	Cultivated Land	Barren Land	Total	% to Total
Slight	178.7	3.4	2.1	185.2	24.8
Moderate	93.3	261.8	21.2	376.3	50.5
Severe	40.2	12.1	128.4	180.7	24.3
Intense	0.4	0.4	2	2.8	0.4
Total	312.6	278.7	153	745	100
Per cent	42.01	37.38	20.61	–	–

A part from these, construction works also help in siltation and degradation. Indiscriminate road construction has done a lot of damage to the forests and fertile land. Each year every kilometer of these roads produce 550 cubic meters of debris by landslides and rock falls[11]. The watershed has total length of roads 457 kilometers which produces on an average 251350 cubic meters of debris, etc, every year which slides down the slopes killing natural vegetation, fertile soil and checking mountain streams as well as down steams to create floods in the plains and siltation.

The physical configuration, biotic pressure of man and animal, developmental activities and managemental problems have lead to the degradation of land resource in the region. As much as 24.7 per cent area is under the threat of degradation. Mainly blank and waste lands have more severe threat because it covers 70 per cent of such area in the region while reserved forests contribute 22 per cent and 8 per cent by cultivable land. The remaining area is also not safe and reveal slight to moderate degradation. Most of the agricultural land *i.e.*, 94 per cent is under moderate erosion. 42 per cent of forested area is under moderate to severe degradation while 58 per cent is slightly degraded in the region.

Natural Resources: Management and Planning

"A healthy mind in a healthy body in a healthy environment" is the basic principle of conservation. Thus every individual has a moral and social obligation to ensure a healthy environment. Environmental issues need to appraise in ethical terms. Such ethics embodies responsibility for the maintenance of the productive capacity of land, air, water and life for their integrity and stability and for the aesthetic qualities of the landscape[12]. The objective of the 'environmental management in the maintenance of environmental quality' for the survival of human species to an environment that not only sustain life but enriches life, harmonizing the work of man and nature for the greater good of all. So, it is to be life supporting, useful and beautiful[13]. It offers a new relationship between man-resource and environment which view the production consumption, function in its wider perspectives for the judicious harvesting of the natural resources. Its aim is to curb the 'misuse' prevent the 'abuse', regulate the 'non-use', foster the 'reuse', and 'save 'and 'invest' resources for the 'future use'.

The impact of environment is totally linked with population[14]. The explosion of population during last 150 years has now drowned the global attention. The teeming population pressure on natural resources has caused hunger, poverty and malnutrition. Several organization governments and UNO has put a question mark on human survival. Thus, the environmental management is the biophysical environment as well as the men-environment. (Socio-culture) interface[15]. The strategy of environmental management varies according to the values, needs, goals performances and perception of people. The people of country/region should feel their moral responsibility to conserve and protect the environmental resources for the prosperity.

Therefore, the central theme of natural resources management is the reduction of minimization of the impact of human activities on the environment, thus, an endeavor to avoid the overuse, misuse and abuse of environmental resources.

People should harness the available knowledge for conservation and regeneration of environmental resources because conservation is a kind of management which ensures sustained yield but does not deteriorate. The environmental education should be given priority for the conservation of resource and minimizing the environmental degradation. A national policy should be declared to encourage productive and employable harmony between man and environment and to promote efforts to prevent environment and the biosphere and stimulate the health and welfare of man.

The main problem is free grazing can be solved by prohibiting order against cattle grazing. The cattle should be stall-fed, grasslands and forest would regenerate and produce four to six times more grass. The day and night time dung would be collected and nature for fields would double. The animals would need to be replaced by smaller number of high yielding animals. The number of bullocks should be reduced. They are used only for 10-15 days in a year for polishing. So, the hill bullock is an almost total liability. For rest of the time they trample the over grazed pastures.

If the cattle numbers are reduced and the cultivated area would have to be curtailed proportionately. The cropping pattern should be reformed to minimize the food grain cultivation to the fertile, irrigated valleys only and put the slopes under: fruit orchards, tea, nuts, medicinal plant, castor, floriculture, vegetable, etc. The plough is automatically banned, the bullocks no longer be needed and large scale felting of oak trees for replacement of the wooden plough would be stopped. Prohibition of grazing could then realy be implemented. The pressure on pasture and forest would be reduced and such land will produce more grass for harvesting. The large areas of bare land will be available for afforestation. The erosion would be stopped, forest revenue would increase, and employment would be generated in forest work and would be come a huge tourist recreation area. A workable channel for the disposal of useless and surplus animals should be adopted.

The environment in hills is badly damaged and more palliatives having the effect of freezing the statuesque will not touch the base of problem. The solution, as suggested, is reduction in the area under cultivation to the point where the relationship between cultivated area, cattle numbers required to support such cultivation, and pasture-cum-forest required to support the cattle are in such proportion that the demand is less than the re-generative capacity.

The ecologists recommend complete afforestation of all lands above 1500 meter altitude (ridges and spurs). Consolation of land holding Act, which has not yet been implemented in the hills, must be enforced and is a pre-requisite to any short of reform of land use. The forest rights should be reserved the locals. Understanding among locals/villagers and their active participation alone can lead to the regeneration of forests on common lands and pastures by social fencing.

References

1. Teli, B.L., 1992. Rural landscape management and planning in Western-Nayar Basin. *Unpublished D.Phil,* H.N.B.G.U. Srinagar Garhwal, pp. 60–70.
2. Teli, B.L., 1995. Soil resources: Problems and planning in hills. In: *Eco-degradation of Himalayas: Causes and Remedies.* (Ed.) B.L. Kaul. Vinod Pub., Jammu pp. 308–338.
3. Shome, K.B. and Raychaudhuri, S.P., 1960. Rating of soils of India. In: *Proc. of Nat. Inst. of Sci. of India,* Vol. 26, A (Supple I): 260–289.
4. Zakrzeveska, Barbara, 1967. Trends and method in landform geography. *A.A.A.G.,* 57: 128–165.

5. Leopold, L.B., Walman, M.G. and Miller, J.P., 1969. *Fluvial Process in Geography*. Indian Eurasis Pub. House, New Delhi, pp. 146–522.
6. Monk House, F.J., 1970. *A Dictionary of Geography*. Edward Enrolled, London, p. 113.
7. Leopard, *et al.*, 1969. *Ibid.*
8. Tempany and Grist, 1958. *Introduction to Tropical Agriculture*. Longmans, London.
9. Teli, B.L.,1992. Management and planning of forest resources in the U.P. Himalayas. In: *Himalayan Environment: Man and the Economic Activities*, (Ed.) J.L. Raina. Pointer Pub., Jaipur, pp. 225–269.
10. Trill, 1915–1839. *Garhwal Settlement Report*, Superintendent, Printing and Stationary, Allahabad pauw, EK, 1895: Report on the Tenth Settlement of Garhwal District, Superintendent, Printing Stationary, Allahbad.
11. Bahuguna, S.L., 1978. Hill development in national context. In: *Environmental Regression in Himalaya: Concepts and Strategies*. (Ed.) J. Singh, Nainital, p. 287.
12. Arivill, Robert, 1969. *Man and Environment*. Penguin Pub., New Delhi, pp. 229–230.
13. Kates, R.E. and Hufschmidt, M.M. (Ed.), 1969. *Comprehensive Environmental Planning in Regional Planning*. Praeger, New York p. 2.
14. Ehrich, P.R., Ehrlich, A.H. and Hare, F.K., 1972. *Population, Resources and Environment*. SanFracisco, pp. 3–4.

Chapter 20

Climate Change Research: Status and Methodology

R.B. Singh and Rajesh Kumar*

Department of Geography,
Delhi School of Economics, University of Delhi, Delhi– 110 007

"Climate change threatens to exacerbate existing vulnerabilities and further entrench development disparities. Those with the least, stand to suffer the most. Thus with regional changes and impacts already being observed, the need for adaptive response measures is imperative. For the poor and other vulnerable people, the need is urgent".

International Institute of Sustainable Development (IISD)

Introduction

Climate change becomes the most debatable issue in the world today. Climate change refers as a long term shift in the climatic conditions of a particular region. These changes can be analyzed in various forms as change in temperature, rainfall or so many associated phenomena. So climate change means a change of climate, which is attributed directly or indirectly to human activity, that alters the composition of the global atmosphere. This change in addition to natural climate variability observed over comparable time periods.

* Corresponding author: E-mail: rbsgeo@hotmail.com

A change in the variability of the climate is also considered as climate change even if average weather conditions could be as normal. India is known as the most affected and vulnerable country due to climate change in the world. Like other developing countries in India mostly population is actively engaged in activities which are more vulnerable, like agriculture, forestry, livestock and many more. People has lack of proper infrastructure, due to this even simple natural climatic variables turns in a severe climatic scenario. Further the means and capacity in developing countries to adapt to changes in climate are scarce due to low level of human and economic development and a considerable high proportion of poor population in the society (Downing *et al.*, 1997).

Vulnerability to climate change means to assess the adaptive conditions of a society in changing climatic phenomena. According to Second Assessment Report (SAR) of IPCC "vulnerability is the extent to which climate change may damage or harm a system. It is a function of both the sensitivity of a system or structure to climate and the opportunities for adaptation to new conditions". The 2007 Human Development Report (HDR) of the UN Development Programme (UNDP) described "vulnerability" in the context of climate change as "an inability to manage risk without being forced to make choices that compromise human well-being over time."

Conceptual Framework

Climate change means fluctuations in the climatic conditions at a large scale. Climate change not always means a considerable rise in earth's temperature but on the other hand a large scale cooling may also be defined as climate change. We may very often take the two words climate change and global warming as synonymous to each other but this is not so. Result from the paleoarchives shows that rapid and profound shifts in earth's climate occurred repeatedly during the late Pleistocene. Climate Change affects all aspects of life, making rainfall less predictable, changing the character of the seasons, and increasing the likelihood and severity of extreme events such as floods (Gupta *et al.*, 2008). Most of the climatic models predicts an increased frequency of dryness in summers while wetness in winters, means a greater risk of drought in the summers. Apart from increasing dryness in summers another important prediction is intense precipitation and occurrences of floods due to increased water holding capacity of the warmers atmosphere. High temperature affects the water quality and exacerbates the various forms of water pollution like pathogens, Dissolved Organic Carbon (DOC), several pesticides and thermal pollution.

Vulnerability is defined in terms of the changes that have occurred or will occur in the selected variables such as household assets, income, natural resources etc. Identifying climate change vulnerability requires a clear conceptual framework. As indicated by the concept contained in the Third Assessment Report of the IPCC where vulnerability is defined as "The degree to which a system is susceptible to, or unable to cope with the adverse effects of climate change, including climate variability and extremes. Vulnerability is a function of the character, magnitude, and rate of climate variation to which a system is exposed, its sensitivity, and its adaptive capacity" (IPCC, 2001). Vulnerability can thus be defined as a function of exposure, sensitivity, and adaptive capacity, or: Vulnerability = *f* (exposure, sensitivity, adaptive capacity).

In the IPCC report, exposure is defined as "the nature and degree to which a system is exposed to significant climatic variations"; sensitivity is defined as "the degree to which a system is affected, either adversely or beneficially, by climate-related stimuli"; and adaptive capacity is defined as "the ability of a system to adjust to climate change (including climate variability and extremes), to moderate the potential damage from it, to take advantage of its opportunities, or to cope with its consequences".

In mostly developing countries millions of poor people are facing the dearth of climate change. Humanity today facing too many problems ranging from social to political and economic but the problem of climate change is quite different from all of these (Kelley and Adger, 2000). Global warming is indicating that we are overloading the carrying capacity of earth's atmosphere by our economic activities. Stock of GHG's increased to a large. Current concentration of GHG's has reached 380 ppm of CO_2 equivalent exceeding the natural range of 650,000 years. In the course of 21st century average global temperature could increase by more than 5°C (UNDP, 2007).

People over different parts of the world have different climatic stresses but the common thing among all is that the public concern about changing climate is continuously increasing. But the other aspect is that mostly affected are the poorest people who have lack of resources (Kumar and Singh, 2003). Agriculture is both the major cause and survival of climate change. Methane extraction from paddy fields, application of huge amount of chemical pesticides and insecticides, crop residue burning etc leads to heavy smoke release and environmental degradation. While on the other hand the major impact of global warming and climate change directly effects the world agricultural system (IPCC, 1996) (Figure 20.1).

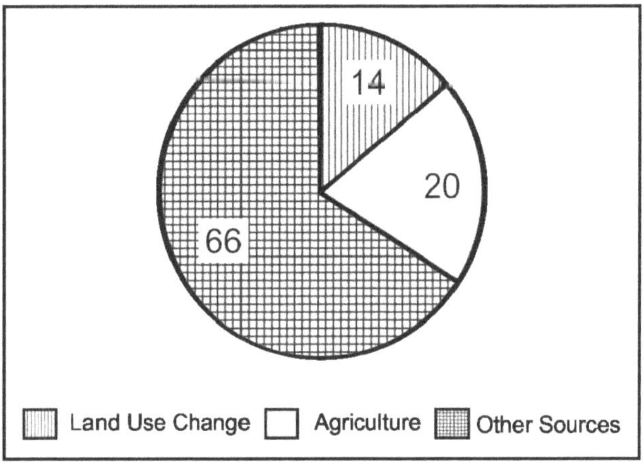

Figure 20.1: Contribution of Different Sectors in Climate Change (per cent) (*Source*: IPCC, 1996)

Emerging Issues

It can be observed that the affected people are largely related to Africa, Latin America or from the under developed Asian countries. According to a report about 262 million people were affected by climatic disasters across the world during 2000 to 2004 and among them about 98 per cent people are from less developed countries.

In the Fourth Assessment of IPCC Report, it is indicated that the vulnerability of a society to climate change was influenced by its development path, physical exposures, distribution of resources, prior stresses, and social and government institutions. All societies have inherent abilities to deal with certain variations in climate, yet adaptive capacities are unevenly distributed both across countries and within societies (Figure 20.2).

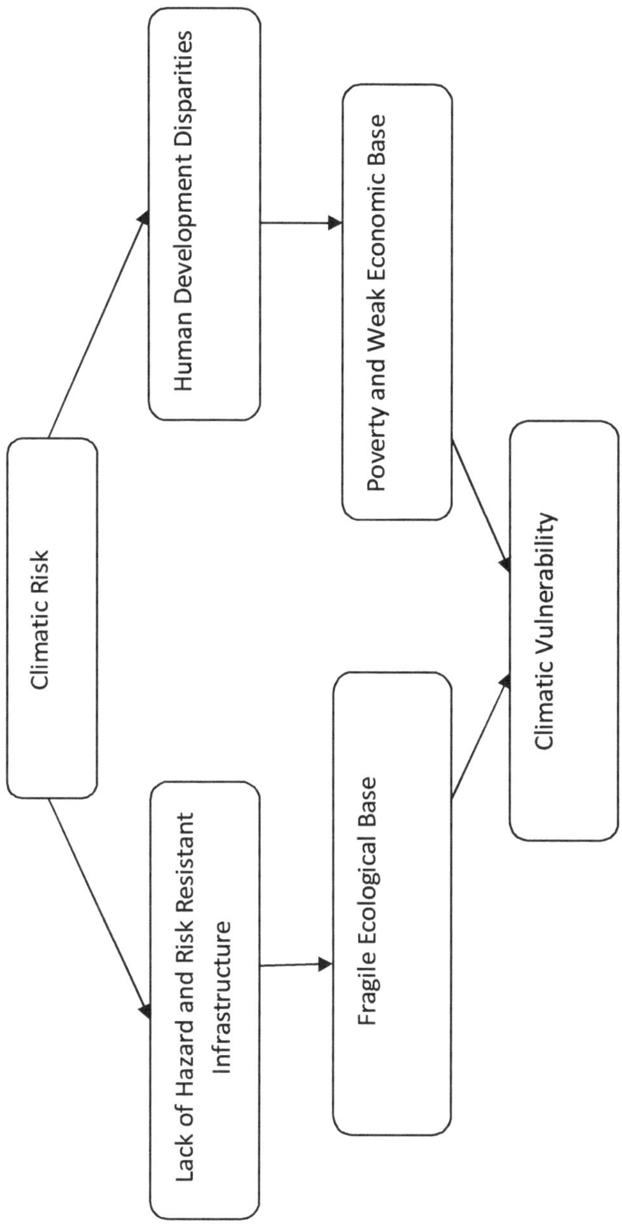

Figure 20.2: Climatic Vulnerability: Conceptual Framework

Statement of Research Problem

Variability in the climate may be a product of natural phenomena but it takes a different picture when this climate change is induced too much by the anthropogenic activities. Human's economic greed has resulted into devastation of natural resources. After industrial revolution the amount of harmful elements in the atmosphere increased in a multiple range. The global mean surface temperature for 2007 is currently estimated at 0.41°C/0.74°F above the 1961-1990 annual average of 14°C/57.20°F. Climate Change represents an additional stress on the systems which are already affected by increasing resource demand, unsustainable management practices and pollution (Kumar, 2006). The impact of these stresses will such that the system shall be unable to provide essential goods and services so as to make a minimum living standard such as adequate food, energy, safe shelter etc.

According to IPCC, the concentration of CO_2 in the atmosphere has increased by 31 per cent over the last 200 years (IPCC, 2001a). If the current trend were to continue this is going to increase by 540 ppm to 970 ppm in 2100 (IPCC, 2001 c). Climate change will have wide-ranging effects on the environment, and on socio-economic and related sectors, including water resources, agriculture and food security, human health, terrestrial ecosystems and biodiversity and coastal zones. Changes in rainfall pattern are likely to lead to severe water shortages and/or flooding. Melting of glaciers can cause flooding and soil erosion (Lal *et al.*, 1998). Rising temperatures will cause shifts in crop growing seasons which affects food security and putting more people at risk from diseases such as malaria and dengue fever. Temperature increase will potentially severely increase rates of extinction for many habitats and species (up to 30 per cent with a 2^{0}C rise in temperature), particularly affected will be arid and semi arid regions.

In semi-arid and arid environments, rainfall is short-lived and often very intense (Leary *et al.*, 2008). Because soils tend to be thin, much of the rainfall runs directly off of the surface, only to infiltrate deeper soils down slope or along river beds. However, in semi-arid and arid areas, where groundwater recharge occurs after flood events, changes in the frequency and magnitude of rainfall events will alter the number of recharge events. Lack of perennial water flow make the situation more critical to already water deficit region.

Assessing the impacts of and vulnerability to climate change and subsequently working out adaptation needs requires good quality information (Metz, 2007). This information includes climate data, such as temperature, rainfall and the frequency of extreme events, and non-climatic data, such as the current situation on the ground for different sectors including water resources, agriculture and food security, human health, terrestrial ecosystems and biodiversity etc. Vulnerability to climate change is the degree to which geophysical, biological and socio-economic systems are susceptible to, and unable to cope with adverse impacts of climate change (Parry, 2007).

Climate change affects the climatic pattern of the particular regions by changing the rainfall and also affects the availability for agriculture in already stressed areas. As per a report the drought hit area of sub-saharan Africa could expand by 60-90 million hectares by 2060. Other similar poor areas are also facing the problem of climate change and resultant decline in agriculture production. Due to declining agriculture production the problems are arising in various poverty elevation programmes so also affect the social system of the affected country also (Sen Roy and Singh, 2002).

Review of Literature

An increasing body of evidences suggest that the global average surface temperature has increased over the 20th century by about 0.6 ± 0.2 degree Celsius (IPCC, 2001 a). Rupa Kumar *et al.*, (2002) indicated that the warming trend over India has been reported to be about 0.57 degree Celsius per 100 years. There are vast regional and sectoral variabilities in India that effect the adaptive capacity of the country to climate change (Bhatt and Sharma, 2002). For example the proportion of irrigated land to total cultivable land is about three fourth in the Punjab on the one hand and even less than a quarter in other states like Orrissa. Being a developing country India is more vulnerable to the effects of climate change due to its high dependence on climate sensitive sectors like agriculture and forestry and its low financial adaptive capacity (IPCC, 2001 b).

Indian farmers are facing extreme climatic events, such as drought, floods, temperature rise/fall etc., every year, with substantial economic loss. Several crops and site-specific studies have been conducted on research farm fields to evaluate the impacts of climate change on crop yields (Aggarwal and Kalra, 1994; Das and Kalra, 1995; Hundal and Kaur, 1997; Kumar and Parikh, 1998; Abrol and Gadgil, 1999; Lal *et al.*, 1999).

Vulnerability depends not only on a system's sensitivity, but also on its ability to adapt to new climatic conditions (Watson *et al.*, 1996). Socio-economic systems typically are more vulnerable in developing countries where economic and institutional circumstances are less favourable (Watson *et al.*, 1996). He denotes that water scarcity, lack of adequate food and other resources in the dry regions make these extreme climatic areas more or less unfavorable for human habitation. Regions having arid, semi arid and hyper arid conditions occupy more than 1/3 of the geographical area of the earth surface. Even with 20th century innovations and advances which made human existence possible in the dry regions they even pose serious obstacles now also. Arid conditions still provides major limitations to the range and extent of human occupations and activities (Kumar *et al.*, 2006).

Local and regional changes (tectonic, climatic, hydrologic, demographic, agricultural, socio -economic etc) combined with global change has raised the concern regarding the sustainability of ecosystem of a particular region. So it is necessary to adopt a multidimensional approach while dealing with the problems like climate change which have wider implication. Recent analysis of rainfall over northern hemisphere land surfaces since mid nineteenth century has indicated significant increase of rainfall in mid latitudes and concurrent decrease in low latitudes over the last 30-40 year (Watson, 1998). In the dry regions of world and particularly in less developed countries, it is the right selection of forage plants and control on land degradation which are the most required mitigation options. By applying these techniques the adverse impact of climate change and global warming can reduced to a significant level (IPCC, 1995).

The increasing frequency of droughts and floods are the most challenging climate change risk to the Indian economy which is required to tackle with utmost care. The geographical coverage of these climatic phenomena is also increasing along with the extreme frequencies. Regional manifestation of global scale climate change processes is very important because of their close link with social and economic prosperity of the affected human population (Singh *et al.*, 2004). The impacts of climate change not only result from the gradual rise in atmospheric temperature but also from increased climatic variability and extreme conditions as floods, droughts and cyclones (UNEP, 2009). The climate change put on risk the infrastructure, health, water resources and environmental conditions in the poor developing countries. To combat the stress of climate change in poor regions it is essential to have more focus on local climatic trends and their causes. Agriculture is the most severely affected sector of the economy as both fertile agricultural land and crop production likely to decrease due to changing climatic conditions. Climate change also have adverse health effects as due to excessive heat stress there is possibility of increase in the number of vector-borne and water-borne disease. Apart from mitigation measures adaptive measure are also important as modifying agricultural practices, improving seed quality, better irrigation systems and education and training programmes. Most of the regional climatic models predicts a 10- 15 per cent increase in monsoon precipitation along with increased frequency of extreme rainfalls in several parts of the country while at the same time there is predictions of 5- 10 per cent decline in rainfall in already drought prone parts of central India. A sharp decline in winter rainfall is also projected which may result to significant reduction in crop yield particularly of wheat.

Ravindarnath *et al.,* (2006) assessed the impact of projected climate change on Indian forest. Under the climate projection for the year 2085, 77 per cent and 68 per cent of the forested grids in India may experience shift in forest types under the A2 (with increasing population, inequity, technological change, more fragmented and heterogeneous world) and B2 (with moderate population growth and economic development, environmental protection and social equity) scenario respectively with serious implication for biodiversity loss.

Kumar *et al.,* (2006) applied a state of the art regional climate modeling system known as Providing Regional Climates for Impact Studies (PRECIS) to develop high resolution climate change scenarios for India. These PRECIS simulations under scenarios of increasing Green House Gas (GHG) concentration and sulphate aerosols show marked increase in both rainfall and temperature towards the end of 21st century. There is no use of all the scientific and technical literature being generated out of continuous high quality research if they do not reach the common people affected by most of these issues (Singh *et al.*, 2004).

The green revolution helped in bringing agricultural prosperity with the introduction of package technology including high quality seeds and fertilizers, better irrigation facilities. But along with growth prosperity this revolution has been caused long lasting environmental impacts in several regions of India as in Haryana. With the excessive use of fertilizers and over irrigation, the ground water in the region becomes contaminated with increased concentration of nitrogen and phosphate, and is hazardous for future crop growth. Due to faulty soil and water management techniques the area under degraded land has been increased causing a lack of fertile agricultural land (Singh, 2000).

Today, climate change is one of the important issues receiving the attention from several quarters in the society. Many scientists have reported the possible consequences as a result of climate change across the world (Shukla *et al.*, 2003). The extent of impact of the projected climate change/variability in the Indian context needs to be addressed because of India's larger geographical area, population and heavy dependence on agriculture for food security. The primary impact of climate change is observed in rural areas where agriculture, animal husbandry and village forestry are severely affected. Apart from environmental impacts climate change also affects the social bondages of rural hinterlands as it increase the rural-urban migration and changes the occupational structure of society.

Natural resources of arid zones, such as timely water supply, fertile soils, vegetation and wildlife, tend to be scarce and the existing resources are easily damaged by changes in precipitation patterns and human actions (Ribot, 1996). A significant increase in the mean minimum and mean maximum temperature by 2-4 ^{0}C depending on the concentration of green house gases in the atmosphere may put large impact on evapotranspiration levels. Due to changing evapotranspiration levels the agriculture and forestry will be largely affected.

Climate change has remained a marginal issue in the immediate agenda of policy making in developing countries in the light of more pressing concerns of underdevelopment like poverty, food security and inadequate access of the masses to health and education services and infrastructure facilities (Singh and Dhillon, 2006). These factors contribute to the vulnerability of developing countries and consequently a higher need for adaptation. Climate has inherent variability and climate change sits on the top of it. Climate change can manifest in gradual change in temperature, precipitation and sea level rise, over and above the natural climatic variability. On the other hand changes in climate could also increase the risk of abrupt and non-linear changes in any ecosystems, which would affect their function, biodiversity and productivity (IPCC, 2001c). Climate change mitigation and adaptation response actions act as added constraints to development, thus transgressing into larger economic development agenda, requiring simultaneous examination of equity and efficiency concerns (Shukla, 1999).

Objectives

The present study includes the following objectives:

1. To analyse the climatic variability in dry regions.
2. To assess the agricultural vulnerability to climate change.
3. To explore the adaptation options and mitigation strategy.

Study Area

The dry regions of Haryana have distinct topographic and geological conditions Figure 20.3). Reverse slope of the region poses difficulty in developing gravitational drainage pattern as canal system. The rainfall pattern of the region has been affected considerably by its continental location and nearness to the subtropical upper-air high pressure of the Thar Desert. The moisture input in the form of rainfall is the major ecological hazard influencing farming and agricultural productivity in the area. On the basis of Thornthwaite's moisture index (1948) region may be categorized into semi arid and dry sub humid climatic regions and collectively called as dry region of Haryana. Districts of Mahendergarh, Bhiwani, Hisar, Sirsa and Jind lies in semi arid category while the district of Fatehabad lies in dry sub humid category. The areas bordering Thar Desert experience hyper-arid climatic conditions. In the present study more emphasis is given to semi arid areas only. So in order to have better understanding and on the basis of available literature and data, six districts of Haryana including Mahendergarh, Bhiwani, Hisar, Sirsa, Fatehabad and Jind are delineated in the study area. The study area has semi arid and dry sub humid characteristics.

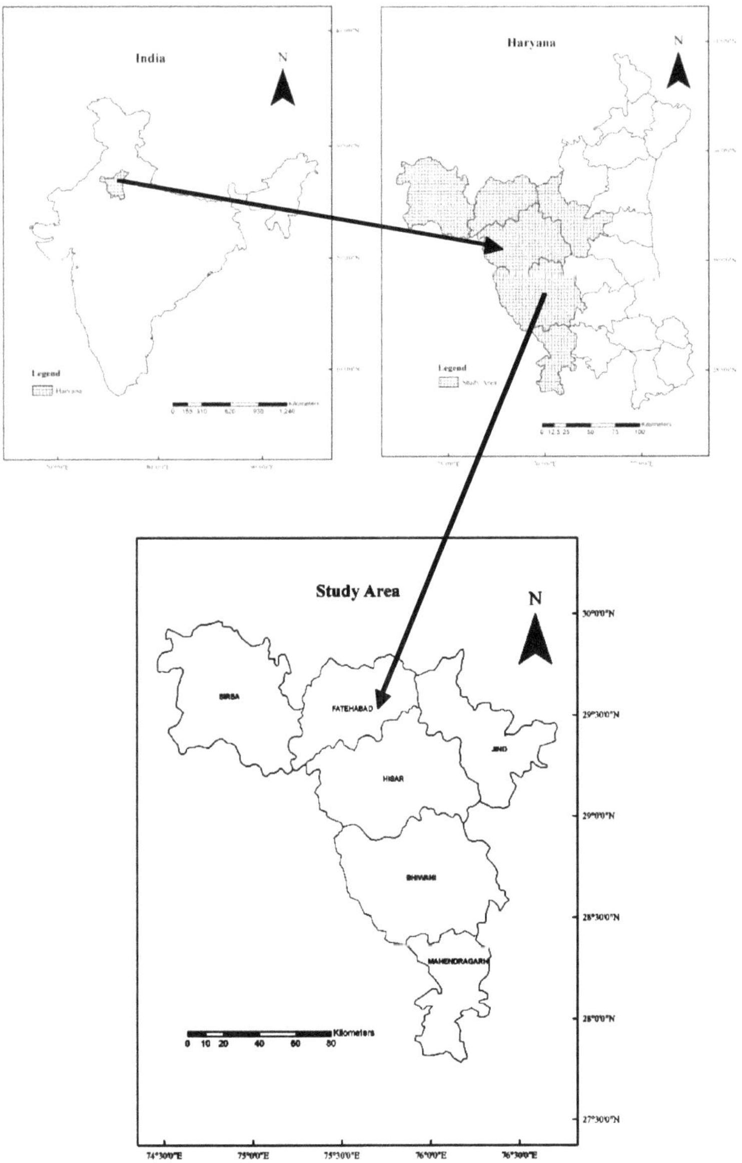

Figure 20.3: Location Map of the Study Area

The total geographical area of the study region is 20,177 sq.km which is about 45.63 per cent of the total geographical area of the state. With a very low area under forest cover almost 3.53 per cent and vast differences in climatic conditions make this region more vulnerable. Due to high population density agricultural sector of the region is under severe pressure.

Research Methodology

Database

Both primary as well as secondary data sources are used in the present study. To identify vulnerability of any region to climate change we have to assess all related social, economic, cultural and environmental information. Data on land holding size, monthly normal rainfall average of 5 years, production of food grains, per cent of irrigated area and decadal population growth rate is collected from Statistical Abstract of Haryana. Demographic data such as population growth, population density, literacy rate have been collected from census of India (2001). District Ground Water Broachers, State Ground Water Profiles published by Central Ground Water Board (CGWB) are the key source of ground water data. Data on surface water is taken from Central Water Commission (CWC) as Yamuna Basin Organization, Delhi and Indus Basin Organization, Chandigarh under Ministry of Water Resources. Climate data like temperature and precipitation of past 60 years from Indian Meteorological Department (IMD). Data related to pest and disease outbreak data from National Centre for Integrated Pest Management (IPM), New Delhi (Figure 20.4).

To analyse forest cover data is taken from Forest Survey of India (FSOI). Application of GIS techniques has been done to transform and manage raw data into some manageable and rational form. Data sets of various forms are analysed, classified and assigned through different index values. Apart from these sources, several other data sources used were used such as CSO, NSSO, Ministry of Agriculture, State Planning Board, Soil and Land Use Survey Ministry of Agriculture. Department of Agriculture, Haryana also provided various data related to soil fertility map, ground water data, Integrated Pest Management (IPM) which is very useful in assessing physical and climatic conditions of the area. Primary questionnaire survey was also conducted to assess people's

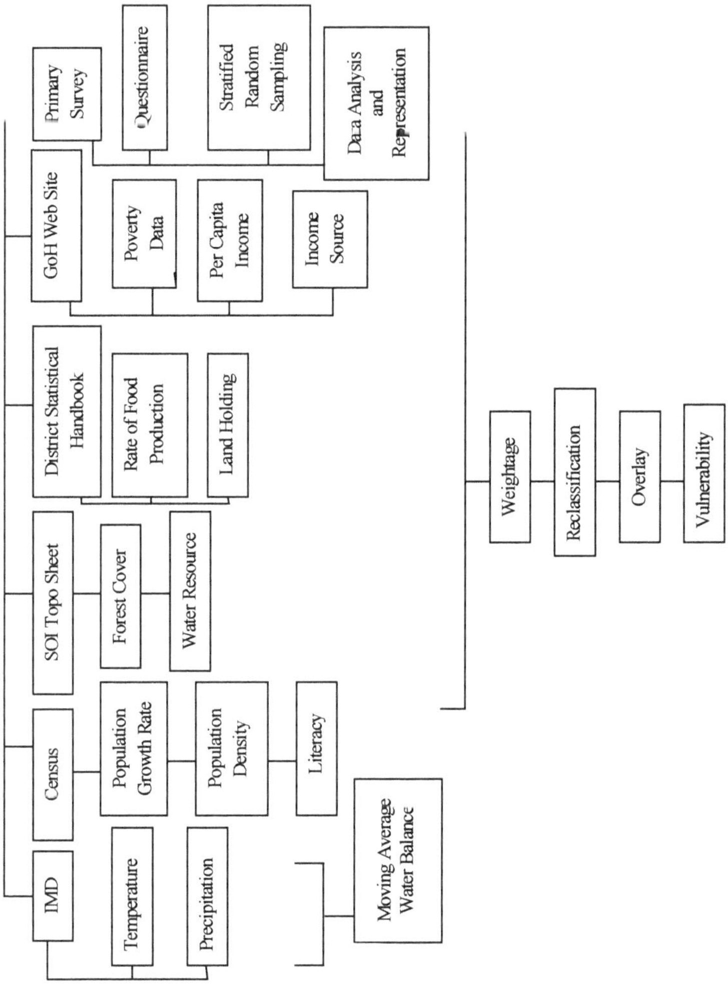

Figure 20.4: Methodological Framework

perception about climate change. Two villages from every district were selected on random basis and from every village 10 respondents were interviewed. So a total of 120 respondents were interviewed. Efforts were being made to include people from diverse geographical localities and representing different cropping systems.

The definitions and delineations of dry regions depend particularly on the purpose of the study and the location of the study area. Scientific definitions have been based on a number of criteria including erosion process, drainage pattern, climatic parameters, plant growth and vegetation types. Whatever criteria are used all schemes involves a consideration of moisture availability at least indirectly through the relationship between precipitation and evaporation. Classification of arid environments which is widely used today was produced on the behalf of UNESCO. Meigs based his classification on the Thornthwaite's 1948 indices of moisture availability (Im).

$$Im = (100\,S - 60\,D)/PE$$

where,

PE: Potential evapotranspiration calculated from meteorological data.

S: Moisture surplus

D: Moisture deficiency

S and D are aggregated on an annual basis from monthly data and taking stored soil moisture in to account. Meigs (1953) identified three types of arid environments as- Semiarid ($-40 < Im < -20$); Arid ($-56 < Im < -40$) and Hyper/Extreme Arid ($Im < -156$).

Data Analysis

The temporal and spatial analysis has been done by using following techniques:

Moving Average (Five years)

This tool is used to find out the trend of temperature and rainfall variation over the years.

Water Balance (FAO and Thornthwaite and Mather)

In this study whenever the monthly rainfall crosses 50 per cent Potential Evapotranspiration (PET), the length of growing period starts and the season gets terminated when the monthly rainfall is lesser than 50 per cent PET. The comparison between this water balance and previous one is helpful in determining the change.

Thornthwaite and Mather (1948), Moisture Index

Moisture index is useful in division of climate of a year into various types. The comparison of moisture indices of each month of this year with previous years is helpful in determining changes in climate for particular months in a year.

Vulnerability Index

The vulnerability index used in this study is based on the index developed by scientists (T.N. Balasubramanian, A. Arivudai Nambi and Diya Paul) working at M.S. Swaminathan Research Foundation, Chennai on 'Vulnerability Assessment and Enhancing the Adaptive Capacity' project under the Climate Change Programme. The index is developed as a statistical tool to assess the existing vulnerability of a region to climate change in terms of the current level of degraded resources. Indicators used in this index represent physical, biological and socio-economic settings of the study region.

The index uses 15 indicators that are sensitive to climate. Each indicator was assigned an optimum value based reflecting the average value of the indicator. The value below or above the chosen optimum mark signifies the level of sensitivity expressed in terms of severe vulnerability, moderate vulnerability and light vulnerability scales.

On the basis of which indicators are more important to indicate vulnerability, these 15 indicators are grouped into 3 categories as First group (Group 1) consists of six indicators, *viz.*, population density, forest cover, water resources/water bodies, rate of food production, area under irrigation and level of literacy which would have severe impact in a region on account of climate change. If the value of the each indicator of the group quantifies as severe (S) according to the assigned scale against each, 100 points/weightage/marks are awarded. Similarly, for moderate (M) and lesser (L) category, 50 and 25 points are awarded respectively. The second group (Group 2) consists of three indicators, viz; rate of population growth, ground water position and poverty per cent, which are considered to have lesser or moderate impact as a result of climate change as compared to the indicators grouped under the first category. For the indicators grouped in the second category, 75 points are assigned for severe vulnerability, and 38 points for moderate vulnerability and 19 points for lesser vulnerability category.

The third group (Group 3) consists of six indicators, *viz.*, per capita income, income source, seasonal dry spell, pest and disease outbreak, land holding size and intensity of soil degradation, which are considered to be relatively less important in terms of their impacts resulting from climate change and in comparison to the chosen indicators under group 1 and 2. The severe vulnerability category in Group 3 carried 50 points, while moderate and lesser category carried 25 and 12.5 points respectively. Thus, the points awarded to each indicator, under the three groups based on values obtained from the collected data serves as the 'Main Factor Effect'. Alternatively points assigned to severe, moderate and lower categories for the indicators given in the first, second and third group could also be derived by the experts based on the expected damage to the system. Based on the direct and outcomes resulting from interaction points, weightage obtained for a particular area, the vulnerability scale of a region/ area could be arrived at.

Major Findings

The analysis of 60 years climatic data reveals the changing trend in the climatic parameters in the dry region of Haryana. Temperature has increased throughout the year, but the most significant is the temperature rise during winter months as in February. There is about 1^{0}C Temperature rise in February from 1948 to 2008. Due to February temperature rise, the *rabi* crops are severely affected in the region. Wheat is the dominant *rabi* crop which experience significant yield reduction due to temperature rise and decrease in rainfall during February and March. Another month taken for vulnerability analysis is of August. As wheat is the dominant affected crop in the rabi season, there are several important crops of kharif as cotton, guar, mung, moath etc which are more vulnerable due to changing climatic patterns. A significant rise in temperature during August is also observed ranging from 0.4 to 0.6°C. The region lies in the ambit of monsoon rainfall and receives most of the rains during 4 months of rainy season from July to September. With the analysis of wet day frequency, it is observed that not only the rainfall regime is shrinking but also, there is significant reduction in number of wet days during the monsoon period. High temperature and reduced rainfall in *kharif* season in August leads to pest and disease outbreak in the region.

Analysis of spatio-temporal maps, makes it quite clear that the northern parts of the dry region, including Hisar, Jind and Sirsa experienced maximum temperature rise of above 0.8°C in February while the Central parts including areas of Bhiwani, Hisar and western Jind districts experienced medium temperature rise in the range of 0.6 to 0.8°C. District Mahendergarh experienced lowest temperature rise in the region in February of below 0.4°C, while Southern areas of Bhiwani district has a temperature rise of 0.4 to 0.6 °C. But the trends of temperature rise are reverse for the month of August in the region. In August maximum temperature rise is recorded in Mahendergarh district of above 0.6°C. The northern areas including districts of Sirsa, Jind and Hisar which experienced highest temperature rise in February has witnessed decrease in temperature in August. These changes are mainly due to the increasing level of Green House Gases (GHG) at the global level but the local factors have also a great role to play in these climatic variations.

Average annual rainfall in the region is calculated about 393.16 mm. With the analysis of 60 years rainfall data, it is clear that rainfall pattern in the region is changing with a significant reduction in total amount of rainfall throughout the year. There is about 40 mm decline in the total annual rainfall which is reduced from about 410 mm in 1948 to 375 mm in 2008. The spatial variation in the rainfall throughout the region makes some pockets more vulnerable to future climatic changes than other regions. The western areas of the region bordering Rajasthan, experience most abundant reduction in rainfall than the eastern parts and northern parts. Maximum reduction in rainfall is experienced in the area surrounding Mahendergarh district with a total reduction of above 60 mm. While the western areas of Hisar, Bhiwani and Sirsa districts experienced a reduction of 20 to 40 mm. The other significant aspect of the rainfall pattern in the region is that rainfall has increased in eastern areas in district Jind. So the Spatio-temporal analysis of rainfall pattern describes that rainfall has decreased both west ward as well as south ward. The major leading factor to declining precipitation is increasing temperature in the region as the high temperature enhances the water holding capacity of the wind and thus delayed in the water condensation.

Wet day frequency has also decreased in the region, the average numbers of wet days in the monsoon period of four months (June to September) were 5.10 per month in 1948 which are now reduced to only 4.68 per month. While there is increase in the wet day frequency from April to June indicating a shifting trend in the monsoon regime and increase in the amount of rainfall caused due to local thermal conditions. The intense heat during April to June leads to uplifting of a parcel of air and rainfall. The frequency of cloud cover has also marked a significant change during last 60 years. The increase in cloud cover is experienced in the months of April to June in summers and October to December in winters. A rise in cloud cover in November and December is more hazardous to *rabi* crops and particularly to the wheat crop. Clouds contributes significantly in warming the atmosphere, they allow passage of solar short wave radiation but absorb the terrestrial long wave radiation. During the sowing period of wheat, it requires low temperature and a high temperature due to cloud cover badly effects the crop yield in the region.

The analysis of the agricultural resource base of a region provides important glimpses of baseline information about the present agricultural scenario for the purpose of formulation and implementation of agricultural planning and policies. The percentage of gross area sown under food grain crops to total cropped area in the region is about 61.16 per cent. The cropping pattern is changing and region is approaching towards mono cropping. Wheat is the dominant crop in the region followed by bajra, having a share of 40.81 per cent and 15.28 per cent respectively out of total cropped area. The area under pluses and oilseeds is reducing and transforming in area under cereals. The area under High Yielding Varieties (HYVs) has increased and in districts of Jind and Hisar, total area under HYV is above 96 per cent. The size of land holdings has been reduced and the total number of small holdings is increased which is increasing the vulnerability of small farmers, there is low output from small farms than big farms in comparison to input. The area under small holdings (less than 2 hectares) is only about 18 per cent while the number of holdings in this size group is about 58 per cent. Ground water is an essential resource for sustainable livelihood in the dry regions, but in the study area the condition of ground water is critical in terms of both quality as well as quantity. In Mahendergarh district about 24 meter decline in ground water is observed in last 20 years. The analysis of land use and land cover scenario of the region provides a detailed picture of the increasing trend of land degradation in the region due to faulty agricultural practices and mismanagement of water sources mainly in Hisar, Jind and Sirsa districts. These areas are also facing the problem of water logging and alkalinity-salinity in the soil.

The agricultural vulnerability to climate change in a region is attributed to several factors. Though the region lies in the semi arid regime and amount of evapotranspiration exceeds the rate of precipitation leading to a scarcity of water resources The natural factors such as rainfall, temperature, cloud cover and wet day frequency etc affect the agricultural conditions of a region, yet some human factors also have a great role to play. High population pressure overlaps the growth and achievements of the agricultural sector as there is no surplus of food for future dry periods. Existing natural resources also have a great bearing of overpopulation due to inherent law the holding size is continuously decreasing among the rapidly growing population. Level of education affect the implementation policies because educated farmers are more aware about the present climatic stress, available resource base and mitigation strategies. Land degradation is another deciding factor of the vulnerability of a region, as more area under degraded land leads to less availability of productive land for food grain production and more consumption for reclaiming the degraded land. Ground water in dry regions is the major source of irrigation as well as for domestic purposes. A decline in ground water makes the region more vulnerable as there is already reduction in rainfall. Due to rainfall decline more pressure lies on ground water sources and a further decline may cause havoc in the region. Another major indicator used is the growth rate of food production. The population is increasing very fast and it is essential to compete the population growth rate with food growth rate. A low food growth rate indicates lack of food for people, malnutrition and low productivity of the people. Low food growth rate may also create a problem of food security.

On the basis of the composite analysis of the indicators used, the study area is categorized into three regions as of high vulnerable, medium vulnerable and low vulnerable. High vulnerable areas include the districts of Mahendergarh and Hisar. In Mahendergarh the problem of ground water depletion makes it less productive.

Due to very deep water extraction the input cost increases and the overall yield goes down. District Mahendergarh has rocky surface and due to stony layer just beneath the upper soil layer, the water holding capacity of the soil reduces. The district has also highest land under degradation in the region. In Hisar the growth rate of food production is very low which makes the area more vulnerable, at a time when most of the authentic reports indicates future decline in crop yield in these regions due to temperature rise.

Moderate vulnerable regions include districts of Bhiwani and Fatehabad. Like Hisar district, the food growth rate in Bhiwani is also very low. Per capita income per day is also low in Bhiwani district, which indicates that a large share of people in the area incapable of maintaining the minimum livelihood standard. High population growth in the area exerts a pressure on the existing natural resource base. Level of literacy is very low in the district in comparison to other districts. In terms of income source of the people the area lies in moderate vulnerability, as more people are dependant on farm income for their livelihood. Poor people are the most bearer of the climatic stress and in Fatehabad district poverty level is very high.

Low vulnerable areas include districts of Jind and Sirsa. These two districts have ample water resources of both surface water and ground water. The vast canal system provides easy availability of quality water. The area under irrigation is high and rate of food production is also good. The area under degraded land is low and thus provides more area for food grain production. Literacy rate in the districts is significant and people are more aware about mitigation and adaptation strategies to climate change.

The increasing population pressure and decreasing agricultural land, along with deteriorating water quantity and quality leads to stagnant agricultural growth rate in the region. There are significant changes in rainfall pattern and the duration of rainfall is shrinking, which has a significant impact on shifting cropping seasons in the region. The productivity of the soils is deteriorating and a significant depletion in ground water can be observed in the region. Though the groundwater table in the region is depleting rapidly the number of electric tube wells is increasing day by day. The electric tube wells are environmentally good, but are capable of extracting water from more deep and thus contributing further water depletion. The level of nutrients in soils is declining and the declining nutrients have adverse health effects on human being as well as livestock. The soil of the region has low water retention capacity which lowers the production. The amount of chemical fertilizers use has increased manifold particularly after the green revolution and package technology.

Adaptation and mitigation policies in a region are complementary to each other to cope with the dearth of climate change. In dry regions water is the prime resource, so the practice of water harvesting needs to be encouraged, particularly in Bhiwani and Mahendergarh districts. Traditional water bodies as *Johads* has strong bondages, from the past with economy and culture of the traditional societies, now in the era of changing climatic conditions, these water bodies needs to be revived. Due to increasing duration of temperature and rise in winter temperature, it is required to grow short duration crops such as of wheat which is ready to harvest before the arrival of early warm periods. Several pockets in Sirsa district such as Nathusari, Baragudha and Dabwali block are facing acute water quality problems. Optimum and judicious use of water resources is the need of hour in dry regions. By applying lift irrigation, sprinkler irrigation and drip irrigation methods, the efficiency of available water resource should be increased several times. Excessive use of chemical fertilizers makes the soil nutrients imbalance, leading to soil degradation. During 1990s the use of chemical fertilizers was negligible, most of the farmers used only organic fertilizers such as *guari khad* and remaining farmers used amount of below 50 kg per acre fertilizers. Now above 40 per cent farmers use above 300 kg per acre of chemical fertilizers. So chemical fertilizers should be replaced by organic fertilizers such as green manures. There is more risk of drought and resultant crop yield loss in several regions of Bhiwani such as Loharu, Tosham and Bhadhra tehsils. Local vegetation which has attributes of coping the climatic stresses is required to conserve. Other adaptation strategies include integrated pest and disease management, maintain soil fertility by avoiding crop residue burning, encourage crop diversity, better livestock management and spreading awareness by education, research etc.

Organization of the Study

The present study is divided into seven chapters. The first chapter deals with the introduction incorporating issues and conceptual framework, literature review, objectives and methodology. Second chapter provides information about geographical background of the study area, its climate, population, soil, drainage etc. Climatic variability of the study region is discussed in chapter three. Causes of variability, extent of variability in terms of rainfall and temperature are measured by using 60 years meteorological data. To assess the agricultural vulnerability of any region, the study considers the existing agricultural resource base which might be going to vulnerable due to changing climatic conditions. The fourth chapter deals with assessing the agriculture resource base including cropping intensity, irrigation intensity, ground water quantity and quality etc. Agriculture vulnerability is assessed in chapter five by applying the vulnerability index analysis. Different indicators as relating to food growth rate, population density and growth rate, literacy level, level of land degradation etc are used to prepare vulnerability index.

References

Abrol, Y. P. and S. Gadgil (1999). Rice in a variable climate. APC Publishing House, New Delhi.

Aggarwal, P. K. and N. Kalra (1994). Analysing constraints limiting crop productivity: New opportunities using crop growth modeling. In: Deb, D. L (Ed.), Natural resource management for sustainable agricultural and environment. Angkor Publishers, New Delhi, 315–332.

Bhatt, J. R. and Subodh Sharma (2002). Impacts of climate change on India and climate change related activities. In: Shukla, P. R., Sharma, Subodh K. and Ramana, P. V. (Eds.), Climate change and India: Issues, concerns and opportunities. Tata McGraw–Hill Publishing Company, New Delhi.

Das, D. K. and N. Kalra (1995). Adjustments to weather variation through cropping systems and fertilizer use. *Fertilizer News*, 40 (5): 11–12.

Downing, T., Ringius, L., Hulme, M. and D. Waughray (1997). Adapting to climate change in Africa. *Mitigation and Adaptation Strategies for Global Change* 2 (1): 19 – 44.

Gupta, K. R., Jankowska, M., Anna, B. and Parsenjit Maiti (2008).Global environment: Problems and policies, Volume 3. Atlantic Publishers and Distributors, New Delhi.

Hundal, S. S. and P. Kaur (1997). Applications of CERES–Wheat model to yield prediction in the irrigated plains of the Indian Punjab. *Journal of Agricultural Science,* 129: 13–18.

IPCC, (1995). Working Group II to the Second Assessment Report of the Intergovernmental Panel on Climate Change. Cambridge University Press, Cambridge.

IPCC, (1996). Climate change 1995: Impacts, adaptation and mitigation to climate change: Scientific technical analysis. Contribution of Working Group II to Second Assessment Report of the Intergovernmental Panel on Climate Change. Cambridge University Press, Cambridge.

IPCC, (2001 a). Climate Change 2001: The scientific basis. Contribution of Working Group I to the Third Assessment Report of the Intergovernmental Panel on Climate Change (IPCC). Cambridge University Press, Cambridge.

IPCC, (2001 b). Climate Change 2001: Impacts, Adaptation and vulnerability. Contribution of Working Group II to the Third Assessment Report of the Intergovernmental Panel on Climate Change (IPCC). Cambridge University Press, Cambridge.

IPCC, (2001 c). Climate Change 2001: Synthesis report. Contribution of Working Group I, II and III to the Third Assessment Report of the Intergovernmental Panel on Climate Change (IPCC). Cambridge University Press, Cambridge.

Kelly, M. and N. Adger (2000). Theory and practice in assessing vulnerability to climate change and facilitating adaptation. *Climatic Change,* 47: 325 – 352.

Kumar, Bhuvan and R. B. Singh (2003). Urban development and anthropogenic climate change. Manak Publications, New Delhi.

Kumar, H. D. (2006). Global climate change: Insights, impacts and concerns. Vitasta Publishing, New Delhi.

Kumar, K. and L. Parikh (1998). Climate change impacts on India agriculture: The Recardian approach. In: Dinar, A., Mendelsohn, R., Everson, Parika, J., Sanghi, A., Kumar, K., Mckinsey, J. and Lonergan, S (Eds.), Measuring the impact of climate change on Indian agriculture. World Bank Technical Paper No. 402. The World Bank, Washington, D.C.

Kumar, Rupa., Sahai, A. K., Kumar, K., Patwardhan, S. K., Mishra, S. K., Revadekar, P. K., kamala, J. V. and G. B. Pant (2006). High resolution climate change scenarios for India for the 21st century. *Current Science,* 90: 334–345.

Kumar, Rupa., Kumar, K. K., Ashrit R. G., Patwardhan, S. K. and G. B. Pant (2002). Climate change in India: Observations and model projections. In: Shukla, P. R., Sharma, S. and P. V. Ramana (Eds.) Climate change and India: Issues, concerns and opportunities. Tata McGraw–Hill Publishing Company, New Delhi.

Lal, M., Singh, K. K., Srinivasan, G., Rathore, L. S. and D. Naidu (1999). Growth and yield response of soybean in Madhya Pradesh, India to climate variability and change. *Agriculture and Forest Meterology* 93: 53– 70.

Lal, M., Whettori, P. H., Pittodi, A. B. and B. Chakraborty (1998). The greenhouse gas induced climate change over the Indian sub–continent as projected by GCM model experiments. *Terrestrial Atmospheric and Oceanic Sciences* (TAO), 9 (4): 663–669.

Leary, N., Adejuwon, J., Barros, V., Burton, I., Kulkarni, J. and R. Lasco (2008). Climate change and adaptation. Earthscan Publications, Sterling.

Metz, Bert (2007). Climate Change 2007: Mitigation of climate change. Contribution of Working Group III to the Fourth Assessment Report of the IPCC, Cambridge University Press, Cambridge.

Parry, Martin L (2007) Climate change 2007: Impacts, adaptation and vulnerability. Contribution of Working Group II to the Fourth Assessment Report of the IPCC. Cambridge University Press, Cambridge.

Ravindranath, N. H., Joshi, N.V., Sukumar, R. and A. Saxena (2006). Impacts of climate change on forests in India. *Current Science,* 90: 354–361.

Ribot, Jesse C. (Ed.) (1996). Climate variability, climate change and social vulnerability in the semi arid tropics. Cambridge University Press, Cambridge.

Sahu, N., Yamashiki, Y. And Takara, K. And R.B.Singh (2011). An Observation on the Relationship between Climate Variability

Modes and River Discharges of the Citarum Basin, Indonesia, Annuals of Disas.Prev.Res. Inst., Kyoto University (Online Jl.), No.54B: 49–55.

Sen Roy, S. and R. B. Singh (2002). Climate variability, extreme events and agriculture productivity in mountain regions. Oxford and IBH Publication, New Delhi.

Sen Roy, Singh, R.B. and Manoj Kumar (2011). An Analysis of Local Spatial Temperature Patterns in the Delhi Metropolitan Area, Physical Geography, Vol. 32: 114–38.

Shukla, P. R. (1999). Justice, equity and efficiency in climate change: A developing country perspective. In: Ference, T (Ed.), Fair weather? equity concerns in climate change. Earthscan Publications, London.

Shukla, P. R., Sharma, S. K., Ravindranath, N. H., Garg, A. and S. Bhattacharya (2003). Climate change and India: Vulnerability assessment and adaptation. University Press, Hyderabad.

Singh, Jasbir and S. S. Dhillon (2006). Agricultural geography. Tata McGraw–Hill, New Delhi.

Singh, R. B. (2000). Environmental consequences of agricultural development: a case study from the green revolution state of Haryana, India. *Agriculture, Ecosystem and Environment*, 82(1–3), 97–103, December. http://www.sciencedirect.com/science?_ob=articleURL

Singh, S., Rao, V. V. M. and Diwan Singh (2004). Climatic variability on local/region scale in India. *Geographical Review of India*, 65 (2)

United Nations Development Programme (2007). Fighting climate change: Human solidarity in a divided world, Human Development Report 2007/2008.

United Nations Environment Programme (2009). South Asia environment outlook. UNEP, SAARC and Development Alternative (DA).

Watson, R. T., Zinyoera, M. C. and R. H. Moss (1996). Climate change 1995: Impacts, adaptations and mitigation of climate change: Scientific–technical analysis. Contribution of Working Group II to the Second Assessment Report of the Intergovernmental

Panel on Climate Change, Cambridge University Press, Cambridge.

Watson, R. T., Robert, Zinyowera, M. and M. H. Richard (1998). The Regional impact of climate change– an assessment of vulnerability. A special report of IPCC Working Group II. Cambridge University Press, Cambridge.

Chapter 21

Diversity of Aquatic Animals in National Chambal Sanctuary, Madhya Pradesh

R.K. Sharma[1], S.C. Bhadoria[1], B.S. Rathore[2] and N. Dasgupta[3]*

[1]*National Chambal Sanctuary, Morena, Madhya Pradesh*
[2]*Department of Geography, Govt. P.G. College, Morena, Madhya Pradesh*
[3]*Conservation Biology Unit, Jiwaji University, Gwalior, Madhya Pradesh*

Introduction

Fresh water and its biodiversity are of vital importance to all life and to nearly any kind of human activity like industrial production, navigation, domestic water requirements, waste assimilation, health, and especially food production (Postel and Carpenter 1997; Gleick 2001). Most of the remaining freshwater wetlands and rivers are currently managed for human uses (Millennium Ecosystem Assessment, 2005) and are being altered from their natural state through changes in hydrology and surrounding land use or extensive resource mining. Surface and ground water extraction has rendered the aquatic biodiversity on

* Corresponding author: E-mail: rksharma_ncs@yahoo.com

the verge of extinction (Dudgeon, 2000; Pusey and Arthington, 2003). Many authors have studied the importance, in terms of values of biodiversity regarding species as well as ecosystems. Total net primary production (Costanza *et al.* 1997); resource usage of freshwater systems (Martinez-Alier, 1995) and ecosystem services associated with biodiversity (Heal 2000) have been studied by many authors.

Freshwater ecosystems are very diverse, complex and dynamic (Ricklefs and Miller, 2000). From being permanently flooded to periodically saturated soil, these are among the richest ecosystems and support riparian and aquatic vegetation, aquatic insects, fishes, mammals and aquatic birds and perform various ecological services (Dudgeon *et al.*, 2006). The structure and balance of these ecosystems are influenced by the physico-chemical feature of water, movement (precipitation, surface and sub-surface flow, direction and kinetic energy of water) and hydro-period (Poff *et al.*, 2010). A healthy ecosystem results from the sum of individual responses of organisms stimulated by the elements in the environment. A greater degree of species or biological diversity of an ecosystem may contribute to greater resilience of an ecosystem, because there are more species present at a location to respond to change and thus absorb or reduce its effects (Levy and Gopalakrishnan, 2010). This reduces the effect before the ecosystem's structure is fundamentally changed to a different state which may be unsuitable for survival of aquatic life (Shine and Klemm, 1999). The biological communities of the world are completely dependent upon the edaphic factors in the ecosystem and their habit and habitats are controlled by their interactions (Ricklefs and Miller, 2000). Thus, the disturbances at the biotic and abiotic interface have led to change in the structure and functions of the biological systems and environmental driving forces, and may damage the integrality of ecosystem (Wang, 2010).

Chambal is one of the few Indian rivers that flow from south to north. The Chambal valley between the Vindhyan and Aravalli hill ranges was the cradle of civilization in central India in the historical past. The river finds a mention as Charmanyawati in most of the ancient Sanskrit texts belonging to the Vedic period. Chambal is a perennial river having its origin in the Vindhyan range near Mhow in Madhya Pradesh. It flows in a northeastern direction through Rajasthan state to meet with its major tributary Parbati river near

Pali in Sawai Madhopur district of Rajasthan. Along its northern course it forms the boundary first between the states of Rajasthan and Madhya Pradesh, then between Uttar Pradesh and Madhya Pradesh and joins with river Yamuna at Pachnada in Uttar Pradesh.

The Gharial (*Gavialis gangeticus*), Mugger Crocodile (*Crocodylus palustris*), Gangetic Dolphin (*Platanista gangetica gangetica*) and numerous aquatic migratory birds are found in the Chambal River. The Chambal River within the National Chambal Sanctuary has been surveyed to assess the strength and distribution of its important aquatic vertebrates, ever since the creation of the Sanctuary in 1979. The frequency of these surveys conducted by various workers has been fairly regular (Sharma, 1981; Singh, 1985; Singh and Sharma, 1985; Sharma and Singh, 1986; Rao and Sharma, 1986; Rao and Sharma, 1987; Rao and Singh, 1987; Rao, 1988; Rao, 1990; Hussain, 1992; Rao and Singh, 1993; Rao and Singh, 1995; Hussain and Choudhry, 1997; Sharma *et al.*, 1997; Sharma and Basu, 2004; Sharma, 2006). The results of these surveys have enabled consistent and sustained monitoring of the aquatic animal populations in this Sanctuary

The present chapter deals with the diversity of aquatic animals in Natioanal Chambal Sanctuary, Madhya Pradesh with special reference to the habitat features and threats against the conservation of these animals.

Methodology

Study Site

The Chambal River originates from the summit of Janapav hill of the Vindhyan range at an altitude of 854 m above the msl at 22°27′ N and 75°37′ E in Mhow, district Indore, Madhya Pradesh. The river has a course of 965 km up to its confluence with the Yamuna River in the Etawah district of Uttar Pradesh (Nath, 1989). From the place of its origin the Chambal River flows for some 320 km in a generally northerly direction before entering a deep gorge in Rajasthan at Chaurasigarh, about 96 km upstream of Kota. The deep gorge extends up to Kota and the river then flows for about 226 km in Rajasthan in a north-easterly direction, and then forms the boundary between Madhya Pradesh (M.P.) and Rajasthan for about 252 km. Thereafter, the river forms the boundary between M.P. and Uttar Pradesh (U.P.)

for about 117 km, enters U.P. near Chakar Nagar village and flows for about 40 km before joining river Yamuna.

Study Area

The river segment that was evaluated during the present study stretches from Pali (Rajasthan) to Pachhnada (Uttar Pradesh) (Figure 21.1). The stretch is about 435 km. The Chambal averages 400 m in width while depth ranges from 1 to 26 m. During monsoon the water level rises 10 to 15 m and often spreads to 500 m from either bank. The maximum discharge of the river is 54, 500 m^3/s and the minimum 27,000 m^3/s. Between 1960 and 1972 four multipurpose dams namely Gandhi Sagar, Jawahar Sagar, Ranapratap Sagar and Kota Barage was built on Chambal River which affected its flow considerably (Hussain, 1992).

Method

The Survey was conducted during 10th to 23rd February, 2011. It started daily on 09:00 hrs till 17:00 hrs. The survey was conducted using motor boat and with the help of GPS device (GARMIN 12), Binoculars (12X50), Data Sheet, Field Map, Digital Thermometer, Depth Finder (GARMIN 90), Camera and Range Finder (BUSHNEL X900). The survey is generally conducted according to the weather condition prevailing at the designated time. Sunny conditions are preferred more than cloudy or foggy conditions. Thus, survey, sometimes may got postponed according to the weather condition.

The spatio-temporal distribution of key aquatic fauna like gharial, mugger, river dolphin, turtles and aquatic birds were recorded at geographical co-ordinates. The population of gharial and mugger was estimated by direct sighting method for different age groups according to IUCN criteria. River Dolphin population was also estimated by direct sighting method at deep pools. The boat was made stationary for at least 30 minutes at locations were dolphins are generally located and repeated counts were conducted to find out the actual number.

The hydrological features of the River, ambient and water temperature, habitat features, shoreline and bank characteristics, anthropogenic activities and threats were also recorded.

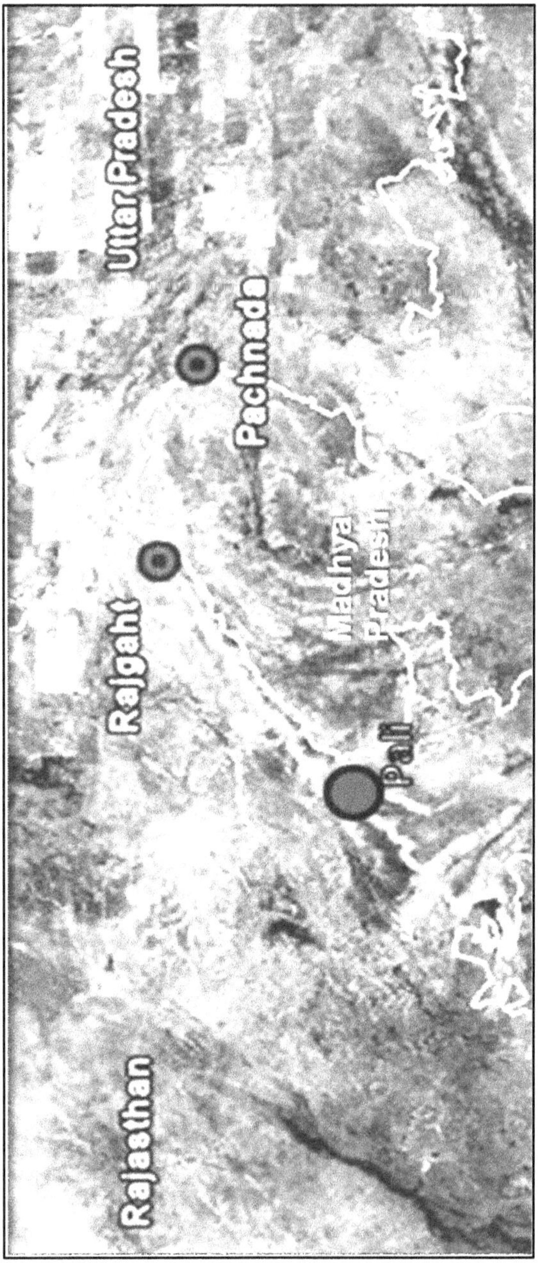

Figure 21.1: Study Site in National Chambal Sanctuary, from Pali to Pachnada

Results and Discussions

A total number of 928 gharial of various size groups (Table 21.1) were sighted in the 435 km stretch of the river. There were 37 males and 360 adult females.

Table 21.1: Sighting of Gharial in National Chambal Sanctuary

Zone Covered (Date)	Adult		Sub-adult (2-2.8 m)	Juveniles (90cm-2m)	Yearlings (60-90 cm)	Hatchlings (30-60 cm)	Total
	Male (>3.5 m)	Female (>2.8 m)					
Pali-Rameshwar	1	6	5	2	1	0	15
Rameshwar-Baroli	0	1	1	0	0	0	2
Baroli-Atar	7	54	9	6	4	5	85
Atar-Saraini	1	6	5	3	1	1	17
Sarsaini-Rajghat	4	38	19	15	17	5	98
Rajghat-Kuthyana	3	54	25	22	16	11	131
Kuthyana-Ratanbasai/ Samauna	1	22	12	9	6	7	57
Ratanbasai/Samauna-Usedghat	2	17	6	6	4	5	40
Usedghat-Ater	3	39	18	15	10	12	97
Ater-Barai	6	55	24	33	16	19	153
Barai-Bindwa	6	63	25	28	16	24	162
Bindwa-Chakarnagar	1	6	4	21	13	8	53
Chakarnagar-Bhareh	0	1	2	5	17	12	37
Total	**36**	**349**	**127**	**170**	**125**	**102**	**928**

The River is densely populated with mugger. The population of mugger was 311 animals (Table 21.2) with 139 adult animals. As it is difficult to identify the sex class by appearance, only size classes has been identified.

Gangetic dolphin the National Aquatic Animal, were sighted mainly at the downstream location (Rajghat to Pachhnada). A total of 65 dolphins of all age group (Figure 21.3) were located (Table 21.3). Due to low water flow and habitat fragmentation the population of dolphin in the Chambal sanctuary is going to decline. The population at Pachhnada was not found during the present survey.

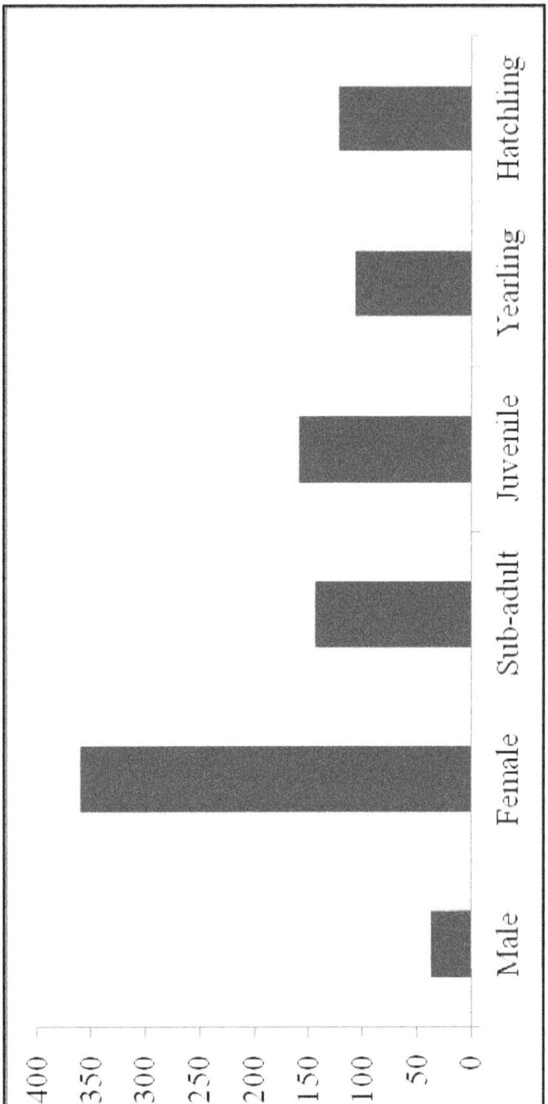

Figure 21.2: Comparative Feature of each Size Class of Gharial Population

Table 21.2: Sighting of Mugger in National Chambal Sanctuary

Zone Covered (Date)	*Adult*	*Sub-adult (2-2.8 m)*	*Juveniles (90cm-2m)*	*Yearlings (60-90cm)*	*Hatchlings (30-60cm)*	*Total*
Pali-Rameshwar	16	10	4	0	0	30
Rameshwar-Baroli	4	3	1	0	0	8
Baroli-Atar	22	15	8	5	1	51
Atar-Saraini	11	3	4	1	1	20
Sarsaini-Rajghat	8	6	4	2	1	21
Rajghat-Kuthyana	16	8	7	3	2	36
Kuthyana-Ratanbasai/ Samauna	14	11	5	2	1	33
Ratanbasai/Samauna-Usedghat	11	8	3	1	0	23
Usedghat-Ater	1	14	6	4	1	46
Ater-Barai	8	7	3	1	0	19
Barai-Bindwa	3	3	2	1	1	10
Bindwa-Chakarnagar	3	2	2	1	0	8
Chakarnagar-Bhareh	2	1	2	1	0	6
Total	139	91	51	22	8	311

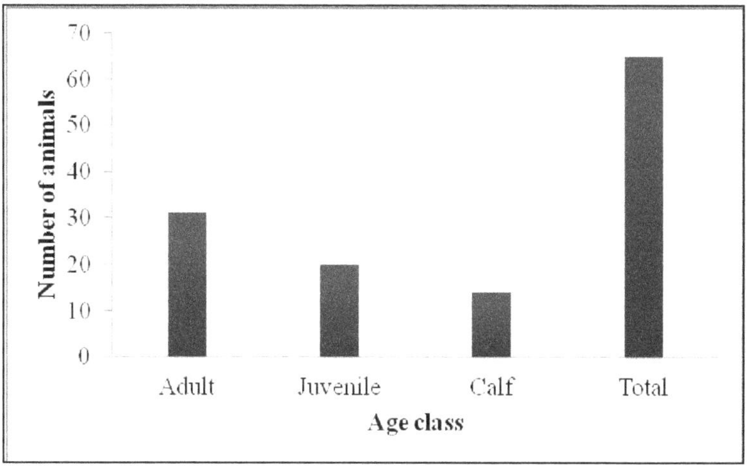

Figure 21.3: Dolphin Population in National Chambal Sanctuary

Table 21.3: Sighting of Gangetic Dolphins in National Chambal Sanctuary

Zone Covered (Date)	*State Boundary*	*Approx. Distance (km)*	*Adult*	*Juveniles*	*Calf*	*Total*
Pali-Rameshwar	MP/Raj.	22	0	0	0	0
Rameshwar-Baroli	MP/Raj.	36	0	0	0	0
Baroli-Atar	MP/Raj.	54	0	0	0	0
Atar-Saraini	MP/Raj	64	0	0	0	0
Sarsaini-Rajghat	MP/Raj.	31	2	1	1	4
Rajghat-Kuthyana	MP/Raj.	30	2	2	1	5
Kuthyana-Ratanbasai/ Samauna	MP/Raj.	25	2	0	1	3
Ratanbasai/Samauna- Usedghat	MP/UP	20	0	1	0	1
Usedghat-Ater	MP/UP	37	2	2	1	5
Ater-BaraiMP/UP	41	5	4	2	11	
Barai-Bindwa	MP/UP	20	9	5	3	17
Bindwa-Chakarnagar	UP	15	5	2	2	9
Chakarnagar-Bhareh	UP	30	4	3	3	10
Total		**425**	**31**	**20**	**14**	**65**

The overall bird population has been depicted in Table 21.4. Some rare species were observed like Common Maganser and Common Shelduck. The highest population was of Pintail which dominated the bird population. Numerous Barheaded geese, Lesser Whistling duck, Red-crested Pochard, Ruddy Shelduck, large cormorants and coots were observed. The absence of Greater Flamingos, very low population of Pelicans and Greylag geese are major concerns. The absence of a greater flock of pelicans can be correlated with the increasing fishing activities at Chakar Nagar to Bhareh. The photographs of some of the animals are presented in Figure 21.4.

The antropogenic activities that are potential threats to propagation of animals are also been observed. Fishing pressure is very high from Barhe (Chambal-Yamuna confluence) to Pachhnada, which may have depleted the prey-base which in turn may affect the dolphin population.

Figure 21.4: The Plate Showing Photographs of Chambal River in Upper and Lower Stretches, Gharial, Mugger with Indian Skimmer, Gharial Hatchlings, Hardshell Turtles, Painted Storks and Interaction with Local Villagers during Survey

Table 21.4: Population Estimates of some Important Avifauna in National Chambal Sanctuary

Avian Species	*Total*
Bar headed Geese	3461
Black bellied Tern	60
Black &Brown headed gull	89
Black Ibis	135
Black Necked Stork	11
Common Crane	5
Common Merganser	14
Common Pochard	38
Common Shelduck	4
Common Teals	86
Coots	1790
Dalmatian and Grey Pelican	16
Darter (Snake bird)	1
Egyptian Vulture	88
Eurasian Wigeon	126
Gadwall	402
Great crested grebe	8
Grey Heron	220
Greylag geese	38
Indian skimmer	224
Large Cormorant	1575
Lesser Whistling Teals	4036
Little Cormorant	315
Little grebe	5
Nakta	543
Northern Shoveller	25
Open-bill Stork	39
Osprey	59
Painted Stork	78
Pintail	4985
Purple Heron	4

Contd...

Table 21.4–Contd...

Avian Species	*Total*
Red-crested Pochard	4705
River Tern	71
Ruddy Shelduck	2404
Small Indian Pratincole	675
Spoon bill	58
Spot-bill Duck	187
Stone plover	53
King Vulture	2
White necked Stork	49
Total	**26684**

Table 21.5: Depth Profile of Gharial and Dolphin Habitats

Sl.No.	*Location*	*Depth (ft)*
1.	Pali	30
2.	Katarni Pura (Baroli)	20
3.	Nadi Gaon	18
4.	D/s Gaughat	23
5.	Atar	21
6.	D/s Rau ka Gaon	123
7.	Hawapura(d/s Chorphandni)	52
8.	Severpali	32
9.	Sarsaini	17
10.	Dang Basai	16.3
11.	Tigri Rithora	50
12.	Puraini	61
13.	Daljit Pura	17.2
16.	Barenda	35
17.	Kherrat	26
18.	Maghera	30
19.	Hatikand ki Dah	152
20.	Barhi d/s	35
21.	Chickney village	44
22.	D/s Chakarnagar	50

Depth also plays an important role for the survival of Gangetic dolphin population which is also decreasing due to water extraction in the Chambal River. These anthropogenic activities may have played a key role in the population decline. The depth profile of some gharial and dolphin habitats are depicted in Table 21.5.

Conclusions

The Chambal River is blessed with a wide variety of aquatic animals. The different terrain features and hydrological regimes provide for diverse habitat features for aquatic animals. Though, anthropogenic activities in terms of fishing, sand mining, stone mining, agricultural practices and grazing pressure was observed throughout the sanctuary. The sand mining activity at the traditional nesting beaches is destroying their viability as nesting and basking ground for gharial and bird species like Indian skimmer. Some of them were on the brink of total destruction. The grazing, agriculture and other human activities degraded the nesting and basking sandy beaches along the River bank. Although fishing activities are totally banned in Chambal River, illegal fishing is found to be a major concern against conservation. Legal mitigative measures, awareness and scientific approach could be the only mean to sustain the natural resources and to protect the habitats from destruction in Natinal Chambal Sanctuary.

Acknowledgements

We are greatful to PCCF Wildlife, Bhopal, Madhya Pradesh for giving permission and support for the study. We are also greatful to DFO, Morena, Madhya Pradesh, for providing all the logistics. We would also like to thank all the field staffs and officials of National Chambal Sanctuary and Deori Gharial Rearing Centre.

References

Costanza, R., d'Arge, R., de Groot, R., Farber, S., Grasso, M.,Hannon, B., Limburg, K., Naeem, S., O'Neill, R.V., Paruelo, J.,Raskinet R.G., Sutton, P. and van den Belt, M. (1997). The value of theworld's ecosystem services and natural capital. *Nature* 387: 253–260.

Dudgeon, D. (2000). Riverine wetlands and biodiversity conservation intropical Asia. Biodiversity in Wetlands: Assessment, Function andConservation (Eds B. Gopal, W.J. Junk and J.A. Davis). BackhuysPublishers, *The Hague.*

Dudgeon, G., Arthington, A. H., Gessner, M.O., Kawabata, Z. and Knowler, D. J. (2006). Freshwater Biodiversity: importance, threats, status and conservation challenges. *Biological Reviews* 81: 163–182.

Gleick, P.H. (2001). Safeguarding our water – making every drop count. *Scientific American* 284: 40–45.

Heal, G. (2000). Valuing ecosystem services. *Ecosystems* 3: 24–30.

Hussain, S.A. and B.C. Choudhry (1997). Status and distribution of Smooth–coated Otter *Lutra perspicillata* in National Chambal Sanctuary. *Biological Conservation* 80: 199–206

Hussain, S.A. (1992). The wild otters of the Chambal. *Sanctuary Asia* 12(5): 25–31.

Levy, J.K. and Gopalakrishnan, C. (2010). Promoting ecological sustainability and community resilience in the US Gulf Coast after the 2010 Deepwater Horizon oil spill. *Journal of Natural Resources Policy Research.* 2(3): 297–315.

Martinez–Alier, J. (1995). The environment as a luxury good or "toopoor to be green"? *Ecological Economics* 13: 1–10.

Millennium Ecosystem Assessment. (2005). Ecosystems and HumanWell–being: Biodiversity Synthesis. World Resources Institute,Washington, DC. USA.

Poff, N. L., Richter, B D., Arthington, A. H., Bunn, S. E., Naiman, R J., Kendy, E., Acremen, M., Jacobson, R. B., Kennen, J. G., Merritt, D. M., O'Keeffe, J. H., Olden, J. D., Rogers, K., Therme. R. E. and Warner, A. (2010). The ecological limits of hydrologic alteration (ELOHA): A new framework for developing regional environmental flow standard. *Freshwater Biology.* 55:147–170

Postel, S.L. and Carpenter, S.R. (1997). Freshwater ecosystem services.In: Nature's Services: Societal Dependence on Natural Ecosystems (edited by G.C. Daily. Island Press, Washington, DC. pp. 195–214.

Pusey, B.J. and Arthington, A.H. (2003). Importance of the riparian zone to the conservation and management of freshwater fish: a review. *Marine and Freshwater Research* 54: 1–16.

Rao, R. J. (1988). Nesting ecology of gharial in the National Chambal Sanctuary. WII, Mimeo pp. 105.

Rao, R. J. and Sharma R. K. (1986). Gharial rehabilitation in Madhya Pradesh, India. *Tiger paper* 13(1):2–4.

Rao, R. J. and Singh L.A.K. (1987) Notes on ecological relationship in basking and nesting site utilisation among *Kachuga* spp. (Reptilia, Chelonia) and *Gavialis gangeticus* (Reptilia, Crocodilia) in National Chambal Sanctuary. *Journal of the Bombay Natural History Society* 84(3): 599–604.

Rao, R. J. (1992). Impact of human activities on wildlife in the National Chambal Sanctuary, India. pp. 171–177. In Tropical ecosystems: Ecology and Management. Ed– by K. P. Singh and J. S. Singh. Wiley Eastern Ltd. New Delhi.

Rao, R.J. (1986). Research on Wildlife in National Chambal Sanctuary and Ghana Bird Sanctuary. pp. 1–26.

Rao, R.J. (1988). Nesting Ecology of the Gharial in National Chambal Sanctuary, Wildlife Institute of India. Mimeo pp.105

Rao, R.J. (1990). Recovered gharial population in the National Chambal Sanctuary. In: Proc. 9th Working Meeting Crocodile Specialists Group, IUCN, Gainesville, Florida (USA), April, 1990. pp. 122–143.

Rao, R.J. and Sharma R.K. (1987). Annual Gharial census in north National Chambal Sanctuary 1985. *Tiger Paper* 14(1): 9–11.

Rao, R.J. and Singh L.A.K. (1993). Communal nesting by gharial *Gavialis gangeticus* (Gmelin) (Reptilia : Crocodilia) in National Chambal Sanctuary. *Journal of the Bombay Natural History Society* 90(1): 17–22.

Rao, R.J. and Singh L.A.K. (1995). Status and conservation of the gharial in India. *Zoo's Print* 10(2): 1–5.

Ricklefs, R.E. and Miller, G.L. (2000). Ecology. 4th ed. W. H. Freeman and Company. NY.

Sharma R.K. (1985). Results of 1981 survey for Gharial (*Gavialis gangeticus*) in North National Chambal Sanctuary. *Journal of the Bombay Natural History Society* 82 (1): 215–218.

Sharma R.K. (2006). Status of Mugger (*Crocodylus palustris*) in protected areas of Madhya Pradesh. *Herpinstance* 4(1): 2–8.

Sharma R.K., Mathur R. and Mishra M.D. (1997). Nesting biology of Gharial (*Gavialis gangeticus*) in National Chambal Sanctuary. *Cobra* 30: 9–12.

Sharma R.K. and Basu D. (2004). Recent Reversals in the population trends in the population of Gharial in the National Chambal Sanctuary in North India; Implications and a suggested strategy for the conservation one of the World's Most endangered Crocodilians. Published in Crocodile Specialist Group, Species Survival Commission, Proceedings of the 17th Working Meeting of the CSG, IUCN. The World Conservation Union, Gland, Switzerland and Cambridge U.K., PP. 180–186.

Sharma R.K. and L.A.K. Singh (1986). Wetland birds in National Chambal Sanctuary. Crocodile Research Centre of Wildlife Institute of India, Hyderabad, 36pp.

Shine, C. and de Klemm (1999). Wetlands, water and the law: Using law to advance wetland conservation and wise use. IUCN, Gland. Switzerland and Cambridge, U.K and Bonn, Germany. Xvi: 330.

Singh, L.A.K. (1985). Gharial population trend in National Chambal Sanctuary with notes on radio–tracking. Govt. of India, Crocodile Research Centre of Wildlife Institute of India, Hyderabad, vii + 166pp.

Singh, L.A.K. and R.K. Sharma (1985). Gangetic Dolphin Platanista gangetica: Observation on habitat and distribution pattern in National Chambal Sanctuary. *Journal of the Bombay Natural History Society* 82: 648–653.

Wang, X. (2010). Research Review of the Ecological Carrying Capacity. *Journal of Sustainable Development* 3(3): 264–265

Chapter 22

Climate Change by Global Warming, Ozone Depletion and Acid Rain: The Alarming Disaster for Mankind

Anurag Paliwal
Assistant Professor,
J.L.N. (P.G.) College, Etah, U.P.

The pollutants of air are not restricted to the earth's surface only. They are carried by wind currents and also ascends up above the atmosphere. Different compounds of the pollutants mainly the gases like oxides of carbon, nitrogen, methane, chloroflurocarbon etc. give rise to secondary environmental problems like acid rains, global warming and destruction of, protective ozone shield in the stratosphere. Together they pose a threat for a serious climatic changes on earth which may spell disaster for the civilization.

Global Warming

Global warming due to "green house effects" induced certain gases like carbon dioxide (CO_2), nitrous oxides (NOx), methane (CH_4) and chloroflurocarbon (CFC-13) in the atmosphere is posing a serious menace to the existence of human life on earth.

As far back as in 1896, Svante Arrhenius of Sweden coin the term "greenhouse effect" and predicted that the burning of fossil fuels would increase the carbon dioxide in atmosphere and lead to warming of worlds climate. In the pre-industrial area (1750-1800 A.D.) CO_2 was recorded at 280 ppmv which by 1896 increased to 300 ppmv and it has topped 350 ppmv level by now. It is currently rising by 0.5 per cent per year. At present rate the CO_2 may increase to 460-560 ppmv by 2400 A.D. Every year nearly 24 billion metric tonnes of CO_2 are released into the biosphere and this is increasing by about 750 million metric tones a year and the rest being absorbed by ocean and vegetation (WWF Report, 1994). CO_2 is responsible for more than half of the warming, chloroflurocarbons (CFC) a quarter, and the remainder is caused by methane and nitrous oxide. Together they will reach equivalent of 560 ppmv of carbon dioxide only by the year, 2030. Methane is 21 times more powerful than carbon dioxide in absorbing heat radiation. The current n1ethanevalue in the atmosphere is 1.72 ppmv more than double the pre-industrial value of about 0.8 ppmv and is increasing at the rate of 0.9 per cent per year, bulk of it coming from rice cultivation, coalmining, domestic ruminants and biomass burning. Molecule for molecule the CFC's are on average, 10,000 times more efficient greenhouse gas than CO_2 for trapping of heat. Carbon dioxide, however reduces atmospheric transparency for the long wave infrared region of the solar spectrum and consequently decreases radiation of the earth's heat back into outer space, thereby intensifying the global warming.

Over the last century the world has warmed by between 0.3°C to 0.6°C. The best estimate is that by 2020 the world will, on average be 1.3°C warmer than now, becoming still warmer by 3°C by 2070 a rate of change unprecedented in the past 10,000 years. A rise of 3°C will make the world hotter than it has been for the last 2 million years. In the past the change of temperature of this magnitude took thousand of years wherein the living organisms could gradually adapt to temperature change, but now it is threatened to be produced in decades due to global warming which would be difficult for many living organisms to cope with. The rate of increase of global mean temperature during the next century would be about per 0.3°C per decade.

The consequences of global warming will he simply disastrous. Sea level would rise by melting of polar ice. One meter rise in sea level could make 200 million people homelsss. With every rise of 1°C plant species will have to move about 90 kilometers polewards to survive. The Arctic tundra may altogether disappear, many parts of earth land mass would submerge under sea water and go out of map. The earth could become a "dust bowl" due to depletion of soil moisture.

Global warming has already begun. Six of the 8 warmest years occurred in the 1980's and the 1990 was the hottest year in history. Oceanic swells, cyclones, tornadoes and storms are increasing in frequency and magnitude. Almost all the 1,196 islands of Maldives suffered huge swells recently, which was not known in the oral history. In 1989, Tavalu, a small island in the Pacific suffered 21 tropical cyclones. In all, 300 Pacific attols may disappear, but will become inhabitable much earlier due to washing by frequent sea storms.

Threat to Human Health from Climate Changes

Even a small increase in humidity and heat could have severe adverse effects on human health. High humidity reduces human's body capacity to discharge heat through the evaporation of sweat. Prolonged cold and wet air can cause worsening of chronic repiratory diseases such as tuberculosis, whooping cough and bronchitis. Prolonged periods of very hot and humid weather can lead to a depression of immune responses in beings. Diarrhea and dysentery develop and diseases like measles, dengue fever and encephalitis occur.

Stratospheric Ozone Depletion

Between 25 and 40Km above the earth surface in the stratosphere lies the protective ozone shield, the natural filter that absorbs and prevents the sun's short wavelength ultraviolet radiation, but for which life on earth would have been impossible. Ozone exists in equilibrium balanced between formation by oxygen and destruction by UV rays. The accumulation of some reactive chemicals in the stratosphere, such as the "oxides of hydrogen", "nitrogen", "chlorine" and "bromine", resulting in the wake of rapid industrialization is posing to accelerate the process of ozone destruction and upset the balance.

Concern for stratospheric ozone depletion started in 1960's due to over emissions of NOx by high flying supersonic aircraft in the altitude of 17 to 20 km. NOx acts as catalyst in ozone destruction. It can remain in the stratosphere upto 1 year. In 1974, it was discovered that the man made chlorofluorocarbons (CFC) used in the refrigeration industries and as propellant and solvents in the aerosol sprays and as the plastic foam blowing agent, was the major destroyer of ozone. CFC can survive for man by years and drift up into the stratosphere slowly in 8 years. In the stratosphere the CFC is destroyed" by UV rays releasing atomic chlorine as free radicals which attacks the ozone molecules breaking it to molecular oxygen and chlorine monoxide (ClO). The ClO further regenerate chlorine radicals and starts in a chain reaction of ozone destruction. Single atom of chlorine is capable of destroying as many as 10,00,000 moelcules of ozone in a chain reaction.

1. UV Radiation→CFC-13= $CFCl_2$ + Cl (Free chlorine radical)
2. Cl+O_3 (Ozone)→O_2+Cl_0 (Chlorine monoxide)-
3. ClO→Cl (free chlorine radical)+ O
4. O_2+O→O_3+Cl=ClO+O_2

In the 1980's it was discovered that "Bromoflurocarbons" (Halons) used as fire extinguishers, Carbon tetrachloride:" used in dry cleaning and "Methyl bromide" used as fumigants in agriculture are also destroying ozone. Methylbromide although have short residency time in the atmosphere (about 2 years), they are 40 times more efficient in destroying ozone than the CFC on a per molecule basis. Some 66,000 tonnes of these chemicals are released into the atmosphere every year.

Damage to ozone layer has been increasing over the Northern Hemisphere where most of the world's people inhabit. According to UNEP every 1 per cent of loss of ozone can result in an extra 50,000 non melanoma skin cancers and 100,000 cases of blindness from Cataracts worldwide. Increased UV-B radiation can cause suppression of the body's immune system and make people vulnerable to infectious diseases. The damage to standing crops forest and vegetation both terrestrial and aquatic could be severe.

Acid Rains

Every year around 100 million metric tonnes of sulphur oxides are released into the global atmosphere. When the atmosphere pollutants like sulphur dioxide, and nitrogen oxides combine with water vapour, sunlight and oxygen in the atmosphere, they create a diluted "soup" of sulphuric and nitric acids which comes down to earth in the form of rain, moist, snow crystals or dry deposition. Investigations have shown that acidic precipitation contains hundreds of organic compounds and many trace metals. Acid fogs in US and other European countries were reported to contain carbonyl compounds, alkyl sulphonate and pesticides, in addition to sulphates and nitrates (UNEP Report, 1992). Acid waters contains increased concentrations of toxic heavy metals such as mercury, aluminium, manganese, lead and zinc. It is aluminum, the most common metal in soil that leaches into lakes. Aluminum toxicity clogs the fish gills and suffocate them to death. Acid rains leaches crucial nutrients such as potassium, magnesium and calcium from soils, depriving trees and other plants of these life supporting elements.

Acid precipitation was first noticed in Sweden and Norway in 1960's, Norway experienced some rainfall that was as acidic as lemon juice. Now acidic precipitation is found in extensive are as of North America and Europe where it is 10 times more acidic than normal. In IJS precipitation as acidic as of vinegar has fallen in some area and 'rain' with a pH value almost equivalent to battery acid once fell in West Virginia. Precipitation containing sulphur dioxide exceeding 1.5 to 2mg/l have been I reported in rapidly developing areas of Asia and South America as well as in Japan and South Africa. There are indications that South-Eastern Brazil, Southern China, South-Western India, Jamaica, North Venezuela and Zambia may experience acidic precipitation in the future if the current trend of use of fossil fuels continues into 21st century.

Acidic deposition poses a threat to various resources on earth *e.g.* lake and aquatic life, forest and vegetation, agriculture and wild life and also to various buildings and monuments. In Germany acid rain is one of the causes of "waldsterben" (Tree death) which affects more than half of the country's forest. Acid deposition is also the main cause of the decline of Swiss forest where 43 per cent of conifers in the alpine regions is dead or dying.

Lastly, If the Human Civilization is to be saved then we have to keep in mind that we have to commercialize the production and utilization of non-polluting energy resources like solar, wind, biogas and hydrogen fuels, and adopt the use of biofertilizers and biopesticides and switch over to organic farming reducing our dependence on agro chemicals.

This planet is ours and we have to save it from extinction.

References

Bakre, Prakash (1995). *Pesticides and Non-target Animals.*

Anonymous (1993). *Action on Ozone.* UNEP Publications. 4: 4–7.

Sinclair, Jain (1991). *Our Planet.* UNEP Publications. 3: 14–15.

Chapter 23

Land-Use and Cropping Patterns in a Gwalior District Village Sankhini: A Case Study

M.S. Sisaudiya[1] *and Balwant Raj Bhadawar*[2]

[1]*Head, Department of Geography, Govt. P.G. Autonomous College, Datia, M.P.*

[2]*Research Scholar, Department of Geography, M.L.B. Govt. Excellent College, Gwalior, M.P.*

Land resource is the most important natural wealth of the country and their utilization is matter of most concern to its people (Radha, 2002). It is necessary for human survival because it provides man with living space with food and with a number of raw materials which are used in the satisfaction of his wants. Man uses land with in several frameworks *i.e.*, physical, social and economic, which often operate together.

Cropping pattern is a particular area unit even at the field level is determined by the fertility of soil, soil texture and structure and by the temperature and moisture components of climate. The Socio-economic variables like the land tenancy, size of holdings, size of fields, irrigation facility, distance from the market, storage facilities and availability of modern input also have a close bearing on the cropping mosaic.

Study Area

The district Gwalior is situated in north-western part of the state Madhya Pradesh. The district covers an area of 4564 sq.km. The district, is divided into three tehsils *viz.* Gird (Gwalior), Pichhore (Dabra) and Bhitarwar and four development blocks *viz.*, Ghatigaon, Murar, Dabra and Bhitarwar. The population of Gwalior district is 2030543 persons as per 2011 census.

Village Sankhini is located on 25°–46′ north latitude and 78°–12′ east longitude in the north east part of Bhitarwar development block. This village covers an area of 1355 hectare and has population of 3697 persons in 2001.

Objective

The main objective of the present paper is to analyse the land use pattern and to recognize the cropping pattern of village Sankhini.

Source of Data and Methodology

The present study is based on primary and secondary data. The primary data was collected by the present authors with the help of Mr. Kripan Singh a farmer of the village in the month of July 2011. The data related to land-use and cropping pattern has been collected from Lekhpal (Patwari) this data is based on agricultural year 2006-07. The land use is based on land-use Gonswara (Bhumi Upyog Gonswara) and cropping pattern data is based on *Kharif* and *Rabi* Jinswar of this village.

Physio-graphically village Sankhini is situated in Sind Parwati doab plain. It is situated 8 km south east of Bhitarwar tahsil H.Q. of Gwalior district. The area of this village is 1355 hectare. It is spread in the Valley of Parwati River and has rich alluvial soil. Total population of the village Sankhini is 4497 persons with 2394 male and 2103 female as per 2011. The literacy rate of this village is 62.4 per cent.

Land-Use Pattern

The land-use statistics (Table 23.1) shows that the village forests share is quite insignificant percentage (2.51 per cent). The net sown area of this village is quite significant as noted in the Table 23.1

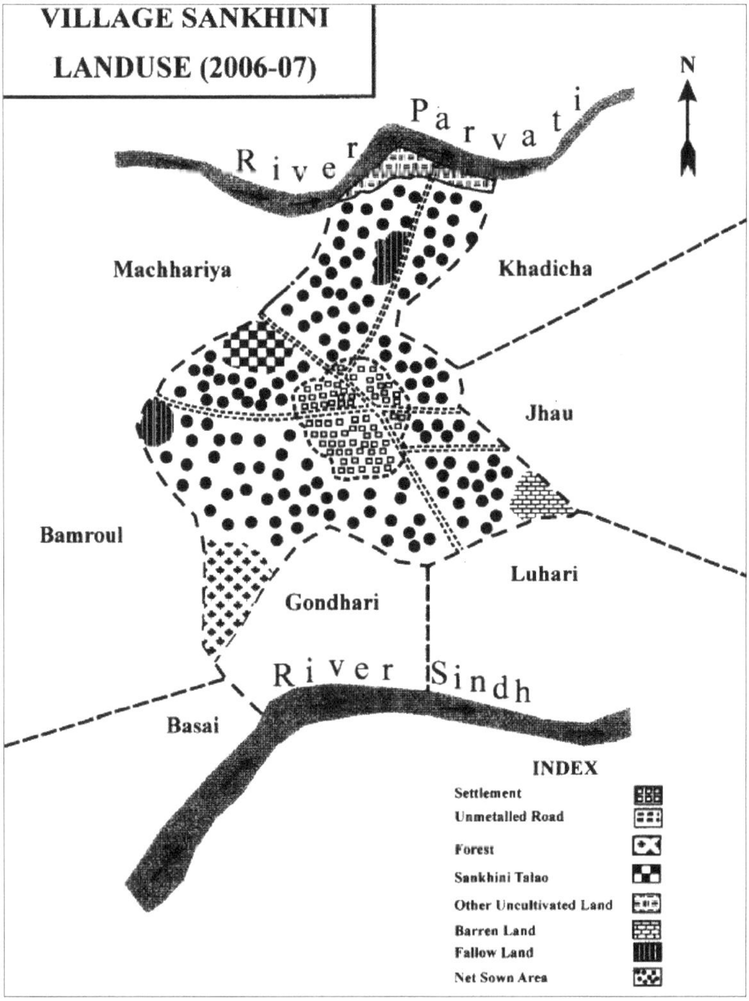

Figure 23.1: Village–Sankhini Land-Use Pattern–2006-07

Table 23.1: Village–Sankhini Land-Use Pattern–2006-07

Sl.No.	*Land-Use*	*Area in Hectare*	*Percentage of Reporting*
1.	Forest	34.01	2.51
2.	Land not available for agriculture other Uncultivated land	201.08	14.84
3.	Land Excluding fallow	36.04	2.66
4.	Barren land	53.44	3.94
5.	Fallow land	50.13	3.70
6.	Net Sown Area	980.30	72.35
	Total	**1355.00**	**100.00**

On account of rich alluvial soil the percentage of net sown area is 72.35 per cent of the total reporting area. Forest area of the village is insignificant *i.e.* 2.51 per cent. Fallow and Barren lands are also in small proportions *i.e.* 3.70 per cent and 3.94 per cent respectively. The land not available for agriculture is 14.84 per cent of the total area.

Cropping Pattern in *Kharif* Season

The area occupied by each crop and their pattern of distribution during the *Kharif* seasons of year 2006-07 has been given in Table 23.2

Table 23.2: Village–Sankhini Kharif Crop Pattern–2006-07

Sl.No.	*Crops*	*Area in Hectare*	*Percentage of Total Kharif Area*	*Per cent of Total Cropped Area*
1.	Rice	96.53	29.70	9.84
2.	Jwar	2.11	0.66	0.21
3.	Bajra	10.69	3.28	1.09
4.	Tuar	2.55	0.78	0.26
5.	Urad	39.84	12.26	4.06
6.	Soyabean	136.12	41.88	13.88
7.	Linseed	16.28	5.00	1.66
8.	Vegetable and other crops	20.92	6.44	2.13
	Total	**325.04**	**100.00**	–

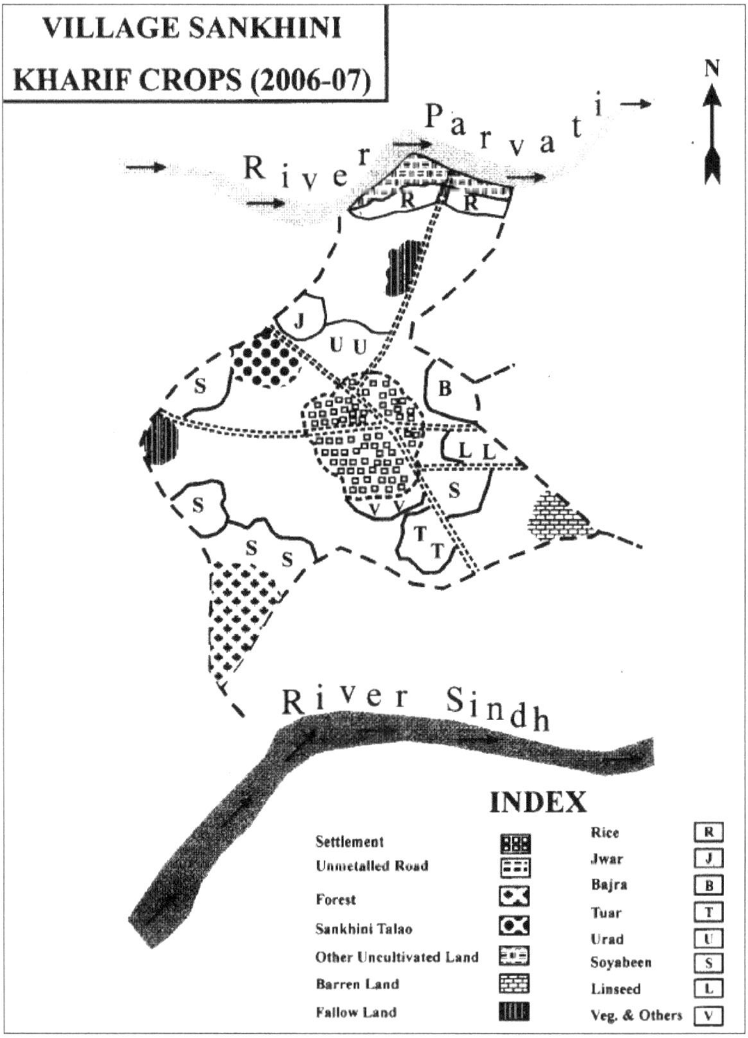

Figure 23.2: Village–Sankhini Kharif Crop Pattern–2006-07

During *Kharif* season 325.04 hectare of land is under crops. In which is 33.16 per cent of total cropped area. Under cereal crops of Kharif like Rice, Jwar and Bajra. 33.63 per cent area is covered pulses occupy 4.32 per cent of total cropped and 13.04 per cent of total Kharif area. Oilseeds *viz.*, Soyabean, Linseed and Groundnut cover an area 152.40 hectare *i.e.*, 15.54 per cent of the total cropped area and 46.88 per cent of total *Kharif* area. Soyabean is the main Kharif crop of this village. It is cultivated in 13.88 per cent of the total cropped area and 41.88 per cent of total *Kharif* area. Rice is grown on 96.53 hectare land of the village Sankhini. Rice is second ranking crop in Kharif season. It is sown in 9.84 per cent of total crop area and 29.70 per cent of *Kharif* season. Jwar and Bajra also cultivated 2.11 per cent and 10.69 per cent of total *Kharif* area respectively. This is noticeable that vegetable and sugarcane are also cultivated in this village occupying 20.92 per cent area of *Kharif* crop.

Cropping Pattern in *Rabi* Season

During Rabi season 655.26 hectare land *i.e.* 66.84 per cent of the total cropped area of the village. In *Rabi* season Wheat, Grain, Mustarq, Peas, Vegetables and others crops are sown in the village. Wheat is the main *Rabi* crop in this village. It occupies an area of 311.23 hectare (47.50 per cent) of total *Rabi* area and 31.75 per cent of total cropped area Mustard is second main *Rabi* crop. It is grown in 155.66 hectare *i.e.* 23.75 per cent of total *rabi* area and 15.88 per cent of total cropped area. Gram is the third main *rabi* crop of this village. It occupies an area of 109.67 hectare (16.73 per cent) of total rabi area and 11.19 per cent of total cropped area of Sankhini village other crops of *rabi* season are Peas, vegetables and others which are cropped 6.44 per cent and 5.58 per cent respectively in this village.

Table 23.3: Village–Sankhini *Rabi* Crop Pattern–2006-07

Sl.No.	*Crops*	*Area in Hectare*	*Percentage of Total Kharif Area*	*Per cent of Total Cropped Area*
1.	Wheat	311.23	47.50	31.75
2.	Gram	109.67	16.73	11.19
3.	Mustard	155.66	23.75	15.88
4.	Peas	42.20	6.44	4.30
5.	Vegetable and other crops	36.60	5.58	3.73
	Total	**655.26**	**100.00**	–

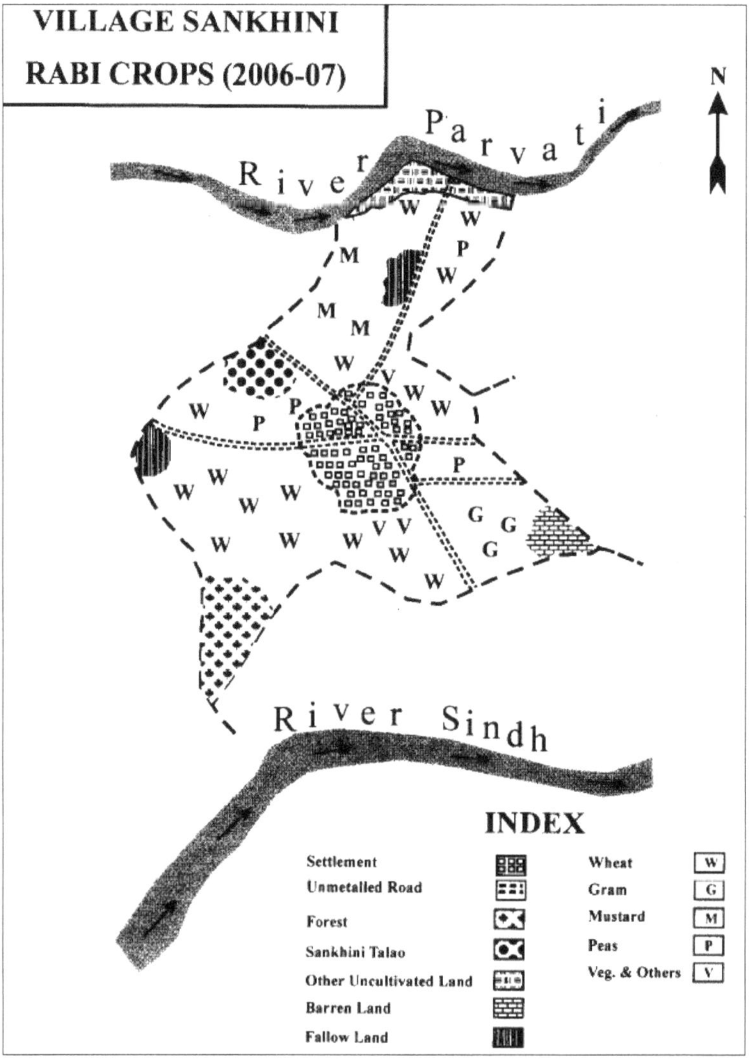

Figure 23.3: Village–Sankhini Rabi Crop Pattern–2006-07

Double Cropping and Irrigation

Double cropping is in this village is due to sufficient irrigational facilities. Tube well, Tank and canal provide sufficient irrigational facilities to this village 401.08 hectare area is double cropped, which is 0.02 per cent of total cropped area. Big bore tube wells, Sankhini talao and canal from Moni Dam provide irrigational facilities to this village. Wheat, Gram and Mustard are irrigated crops in this village.

Conclusion

The land-use patterns and cropping patterns of village Sankhini indicates the medium cum high socio-economic conditions of the farmers of this village. Net sown area of this village is 72.34 per cent which is quite sufficient, which cropping pattern of this village shown healthy agricultural system. Wheat, Mustard, Soyabean, Gram and Rice are major crops. Which are cultivated for their own consumption and to export other parts of the district and state. The farmers of this village and economically sufficient and literate. They use modern agricultural implements *viz.* Tractor, Thressurs and Harvesters and modem methods of cultivations. The village has sufficient irrigational facilities. The farmers use chemical fertilizers D.A.P., Urea, Superphosphate and high yielding variety seeds, agricultural instruments and insecticides.

The village has cooperative societies, cooperative banks and other commercial bank to provide financial assistance to villagers.

The author wish a bright future of the villagers if they continue such agricultural practices and Government look after their needs.

References

Agrawal, G.D. and Bansal, P.C., 1969. *Economic Problems of Indian Agriculture*. Vikas Publication, New Delhi.

Gupta, H.S. and Shukla, Kirti, 1986. Agricultural mechanization in Raipur: Bhilai Axis Region. *Geo-Science*, 4(1): 14–18.

Hussain, Mazid, 2000. *Agricultural Geography*. Rawat Publication, Jaipur.

Jain, C.K., 1988. Agricultural land-use. In: *Patterns of Agricultural Development in Madhya Pradesh*, (Ed.) C.K. Jain. Northern Book Centre, New Delhi, pp. 95–105.

Radha, 2002. Agricultural Development and Planning in Auraiya District: A Geographical Study. *Unpublished Ph.D. Thesis*, Dr. B.R. Ambedkar University, Agra, p. 178.

Roy, S., 1973. Land-use in Upper Mahanadi Basin. *Unpublished Ph.D. Thesis*, Pt. Ravishankar Shukla University, Raipur (C.G.).

Verma, Shalini, 2010. Land-use and cropping patterns in Rajnand Gaon district village Bodal: A case study. *Charmawati Geographical Research Patrika*, 10: 29–35.

Chapter 24

Problem Soils: Their Reclamation and Management

Udai Veer Singh Yadav[1] *and P.K. Kulshershtha*[2]

[1]*Principal,* [2]*Head, Department of Geography, J.L.N. College, Etah, U.P.*

S = SAVE
O = OUR
I = INDIAN
L = LAND

'If the soil is gone, Man must go and everything will be perished.......?

"Soil Erosion is a Disease"

"Soil Conservation is a Cure"

"Soil is a layer of unconsolidated materials at the earth's surface of which has been derived from the rocks."

In India almost 130 million hectares of 45 per cent of total geographical surface area, is affected by serious soil erosion through gorge and gully, shifting cultivated wastelands, sandy areas, deserts and water lodging. Soil erosion by rain and river that takes place in hilly areas landslides and floods, while cutting trees for agricultural implements, firewood and timber; grazing by a large number and other activities, have all fed to the opening of hill-faces to extreme soil erosion. About 90 million hectares of land in India have been affected by water erosion and about 50 million hectares by wind erosion. According to various calculations, about 6000 million tons of good top soil is displaced every year. On the average, about 7 million hectares are affected by floods every year and about 6 million hectares suffer from water-logging. It takes nature 5000 to 1,000 years to build one centimeter of fertile top soil. Yet since Independence, we have paid scant attention to prevent loss of such soil even though 75 per cent of won people are dependent on land for their livelihood. To date we have lost nearly 100, million hectares out of total geographical area of 320 million hectares through erosion by water and other causes.

Problem of Soil

1. Soil Erosion
2. Water logging
3. Alkaline
4. Waste land
5. Expansion of Desert Land
6. Exploitation of land by man
7. Exploitation of Agricultural or land

Types of Soil Erosion in India

There are several causes that lead to various types of India, such as natural factors, like wind and water factors like deforestation, mining etc. The different erosion in India are Normal of Geologic Erosion, Air Erosion, Wind Erosion, Water Erosion, Sheet Erosion, Landslides or Slip Erosion and Stream-Bank Erosion. The causes of destruction of forests and soil erosion, and the nature and severity of the erosion problem vary greatly in different parts of the country. In some areas as in the forests of Assam, Bihar, Odisha and Madhya Pradesh, shifting cultivation. Which is practiced by the tribal people living in these areas is a major cause of destruction of forests.

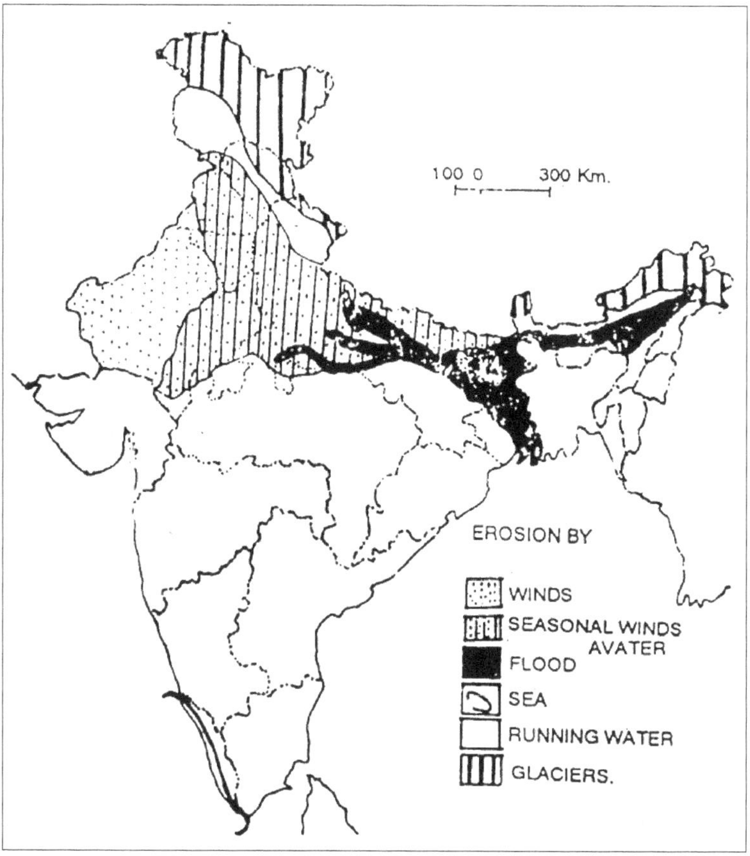

Figure 24.1: Soil Erosion in India by Various Agents

Erosion

E = Eradication

R = Rape (By force)

O = of

S = Soil

I = in

O = our

N = Nation

- ☆ Sheet Erosion
- ☆ Gully Erosion

Soil Conservation

Soil Conservation is a set of methods and procedures which are adopted for prevention of soil being eroded from the earth's surface or becoming chemically altered by overuse of Salinization or acidification. The main methodology of soil conservation are improvement of the choice of vegetative cover, soil erosion prevention, salinity management, to augment the health of beneficial soil organisms and stopping of soil contamination. Soil conservation if a utmost importance for a country like India which is mainly has an agricultural economy. A large part of India's population depends on the soil for their livelihood and hence soil erosion and the-measures of soil conservation taken up to prevent soil erosion are of key concern for the Indian government.

Conservation

C = Creeping Death of Soil

O = of

N = National

S = Soil

E = Engineering Measures

R = Research and Reclamation Method

V = Valley Project Areas

A = A forestation and Agriculture

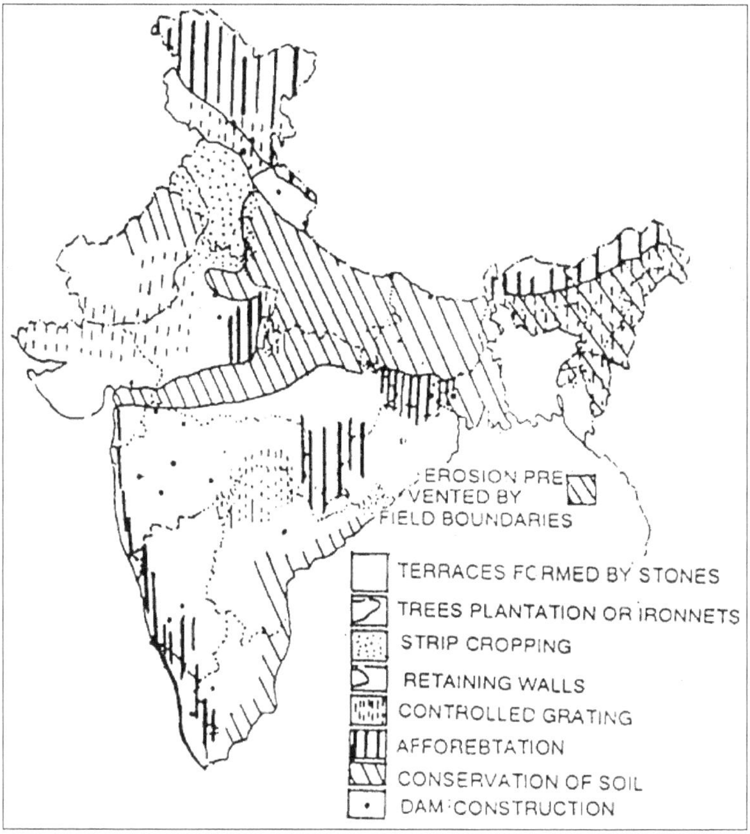

Figure 24.2: Soil Conservation Practices in India

T = Training of Personnel
I = Improvement of Land Use
O = Orientation Programmes
N = Non-Govt. Organization

Types of Soil

Pedalfers

PED = Soil
AL = Almunium
FERS= Iron

Pedocals

PEDO = Soils
Cals = Calcium

Measures for Controlling Erosion and Restoring the Productivity of Eroded Lands can be Divided into Four Classes

1. *Regulation of land use*: This includes all measures for securing such alterations in the existing patterns of land use as are necessary to ensure that the different types of lands are used according to their land use capability *i.e.* the use for which, in view of their physical characteristics. They are best fifted Retiring cultivated lands in highly erodable areas from cultivation and bringing them under forests of grazing; restrictions on or closure of grazing in badly eroded forests or grazing lands, and settlement of shifting cultivators to permanent cultivation are some examples of the types of alterations which are needed.
2. Forestation and preservation of forests by scientific forest management.
3. Improvement of land use practices on farm lands. This includes such measures as plugging along the contours and strip-cropping on sloping lands; proper crop

rotations; application of adequate manures and fertilizers, care of follows and other uncultivated lands.

4. *Engineering measures*: Under this are included construction of bunds and terraces, check dams channels for drainage of surplus water, gully plugging, etc.

A comprehensive programme of soil conservation for an area would include all four types of measures, although the relative importance of the different measures would vary greatly in different areas depending upon the particular conditions of the area.

The Problem of the Rajasthan Desert

Desert and semi-desert conditions prevail over a large area in western India- in Rajasthan and the adjoining areas of Punjab, Patiala and East Punjab States Union (PEPSU), U.P., Saurashtra and Kutch. The planning Commission in its Draft Outline Report drew attention to the reported advance of the desert and encroachment of sand on fertile lands. The Government of India recently appointed an ad hoc Committee of experts of investigate this problem. As a first step a Desert Research Station is being set up at Jodhpur and a pilot scheme for the creation of vegetation belts and the improvement of land use is being taken up. Research on soils, land-use and silvicultural practices would be undertaken at this station.

As a number of states are affected by the advance of the desert, an as successful tacking of the problem would depend in large measure upon co-ordinated action by the various States concerned, the need for inter-State cooperation in this sphere cannot be over-emphasized. We recommend that a coordinated programme of action, indicating clearly measures to be taken in each State should be jointly worked out by the various states concerned, in concerned, in consultation with representatives of the Central Organization.

Organisation

For carrying out the programmes outlined above and generally for formulation and implementation of suitable policies in the fields of land utilization and soil conservation, we recommend the constitution of (a) a Central Land Utilisation and Soil Conservation Organisation at the Centre, and Land Utilisation and Soil

Conservation Board in every State.

The functions of the Board will be

1. Assessment of the soil erosion problem in the State. For this purpose a reconnaissance soil erosion survey should be carried out.
2. Preparation of plans for control of erosion and soil conservation in the State.
3. Drawing up suitable legislation for the execution of improvement of land use and soil conservation programmes.
4. Execution of plans *e.g.* construction of bunds terraces and other works; demonstration of soil conservation practices, aid to cultivators for execution of approved soil conservation programmes, and promoting formation of soil conservation associations. Those of the measures which lie in the sphere of action of other Departments should be undertaken through the regular agencies of the Department concerned. Thus the necessary restriction on grazing feeling etc should be enforced by the forest Department. The presence of representatives of the various Departments concerned on the Board will ensure the co-operation of these Department.
5. Framing suitable programmes for demonstration and research publicity and training of personnel.
6. Supervision and control of Soil Conservation Associations.

By drawing attention to these soils, by emphasizing the ways their properties affect their reclamation and improvement, by discussing some of the lessons learned during the last decades, and by mentioning local solutions to the proper use of these soils, we hope to contribute to a better understanding of the problems encountered and the risks involved when such soils are used for agriculture.

Chapter 25

Cropping Pattern and Crop Combination Analysis for Sustainable Agriculture Development: Buldhana District, Maharashtra

Vijay B. Kharate
Head, Department of Geography,
Arts, Science College, Kamargaon, Dist. Washim, M.S.

ABSTRACT

The present study is an attempt to analyze the agriculture land use pattern and crop combination region, at micro level in Buldhana District. This study is based of secondary data collected from revenue records in both *Kharif* and *Rabi* crops. Agricultural production is influenced by physical, socio-economic technological and organization factors. The crop data has been computed with the help of Weavers (1954) method of crop combination. The study region cover 9661 sq.km land having 1886299 population in District in 2001 censes. Buldhana district is located at the western part of Vidharbha. Fertile region on Purna and Painganga Basin physiographic, rainfall, soil, temperature and drainage influences on agricultural crop combination region. 38 crops have been considered for crop combination on analysis among these cotton, jowar, Bajra,

Oilseeds, Sugarcane fruits crops, pulses, and vegetable crops, are major crops by computing cropping pattern and using Weavers minimum deviation method crops combination of the Buldhana District identified 6 crops combination such type of study represent real situation of cropping pattern in Buldhana District and helps to planners and agricultural scientist for agricultural planning at Tahasil level.

Introduction

The cropping pattern is however, a dynamic concept as it changes over space and time, crops are generally grown in combination in any region. These crops have its relative position in terms of crop combination. Farmers are growing numerous cropping the field rather than single crops. The distributional pattern of crops in any region is an out come of predominance of certain crop or combination. This is in term of emergence of typical crap combination. The crops are generally grown in combinations and its is rare that a particular crop occupies a position of total isolation from other crops in a given a real unit of a given point of time the distribution maps of individual crops are interesting and useful for planners out it is even more important to view the integrated assemblage of the varies crops grown in on a real unit.

In this chapter on attempt have been made to analyses the crop combination regions from the price of view of dominant and distinctive crop regions.

Study Area

Buldhana is the Western most district of the Vidarbha region as well as Amravati Division of the state of Maharashtra. The study region lies between 19°51" and 21°17" North latitude and 75°57" and 76°59" East longitude. North is Bounded by the east Minar district of Madhya Pradesh. While along the East Akola and Amravati district. South West Jalana and Aurangabad Districts. The study are covers an area of 1961 sq. km with a population of 1886299 as per (2001) census. There are two main rivers namely Painganga and Purna. Buldhana plateau a have hill rising to above 500 m. In regard to rainfall, it varies from west to east. The average rainfall in the study area is 798 mm on. The satapudas have shallow, gravelly and stony reddish soils. The alluvial plains of Puma and Buldhana Plateau are covered by deep black soils.

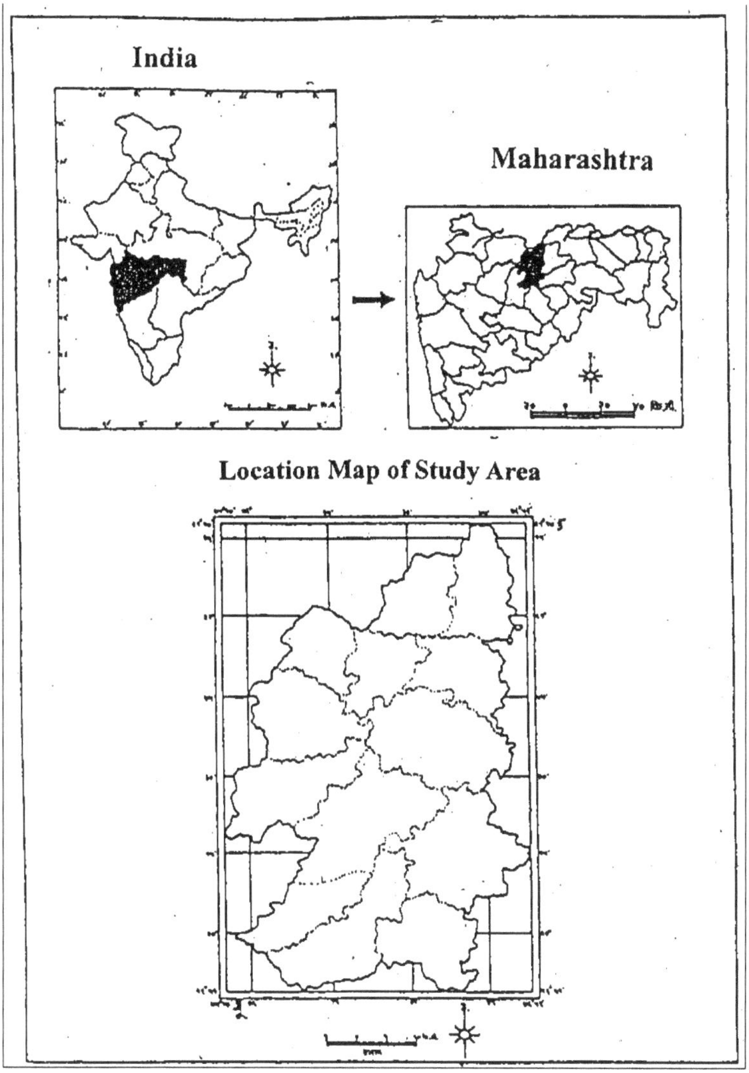

Figure 25.1: Location Map of Study Area

Objectives

The main objectives of the present study are enunciated here under.

1. To attempt the temporal variations in crop combination of Buldhana District, to demarcate the spatial
2. To variations is crop combination at tahasil level.

Methodology and Database

In the present study identify the crop combination region of Buldhana District have been calculated by using ranking method. This is a simple, descriptive, non-quantitative approach to the delineation of crop associations zones. It involves the ranking of individual crops according to their area importance in each component unit. To study the extent of crop combination analysis among the 13 Tahsils of Buldhana' District, Weavers Corp combinations methods have been chosen.

The secondary data have been collected from varies is Govt. Published reports of Buldhana district for the years of 2000 and 2001 to analysis the crop combination regions and cropping pattern. Circle wise crop data is obtained from Talathi records and Panchyat Samiti records in Tahasil lavel.

Exposition

Recently the crop combination, analysis geographical studies has gained momentum and its imp: Is increasing day by day. Combination is in it self and integrative reality that demand definition and distribution analysis and lastly crop combination regions are essential for the construction of still more complete structure of vivid agricultural region.

According to this method for present study in weavers crop combination (minimum deviation method) and statistical calculating from *Kharif* and *Rabi* seasons data (Table 25.1). The Least sum of squared Deviation and Variance and lowest standard Deviation and coefficient of variation formula was selected for this study and analysis from (2000-2001) both (*Kharif* and *Rabi*) seasons variance value (Table 25.1 and Figure 25.2).

Formula my be used as

$$d = Ed^2/n$$

where,

d: Deviation

N: No. of Crops

Ed^2: Deviation from actual magnitude of the deviation eliminated the square root.

Table 25.1

Tahsils	*Name of Crop*	*Crop Combination*	*Coefficient Value*
Nadura, Malkapur, Motala D. Raja	Cotton, Jawar	Two Crop combination	40
S. Raja	Cotton, Wheat, Jawar	Three crop combination	51
Shegaon	Cotton, Jawar, Pulses, Wheat	Four crop combination	59
Jalgaon Jamod Khamgaon Mehkar, Lonar Sangarmpur	Cotton, Jawar, Wheat Pulses, Oilseeds	Five crop combination	74
Chikheli Buldhana	Vegetable Soyabeen Cotton Jawar, Wheat Pulses, Oilseeds	Six crop combination	85

Source: Computed by Author.

As per weaver's method during (2000-2001) other was no mono culture two crop combinations and three crop combinations whereas four, combination, six combinations were present only in Nadura, Melkapur, Mothala, D. Raja Tahsils.

The Buldhana district has identified six crop combination regions. It includes Cotton, Jawar, Tur, Soybean, Wheat, Pulses and oil seeds.

The Tehasil level crop combination to selection crops Buldhana district has completed in this study (Table 258.1 and Figure 25.2).

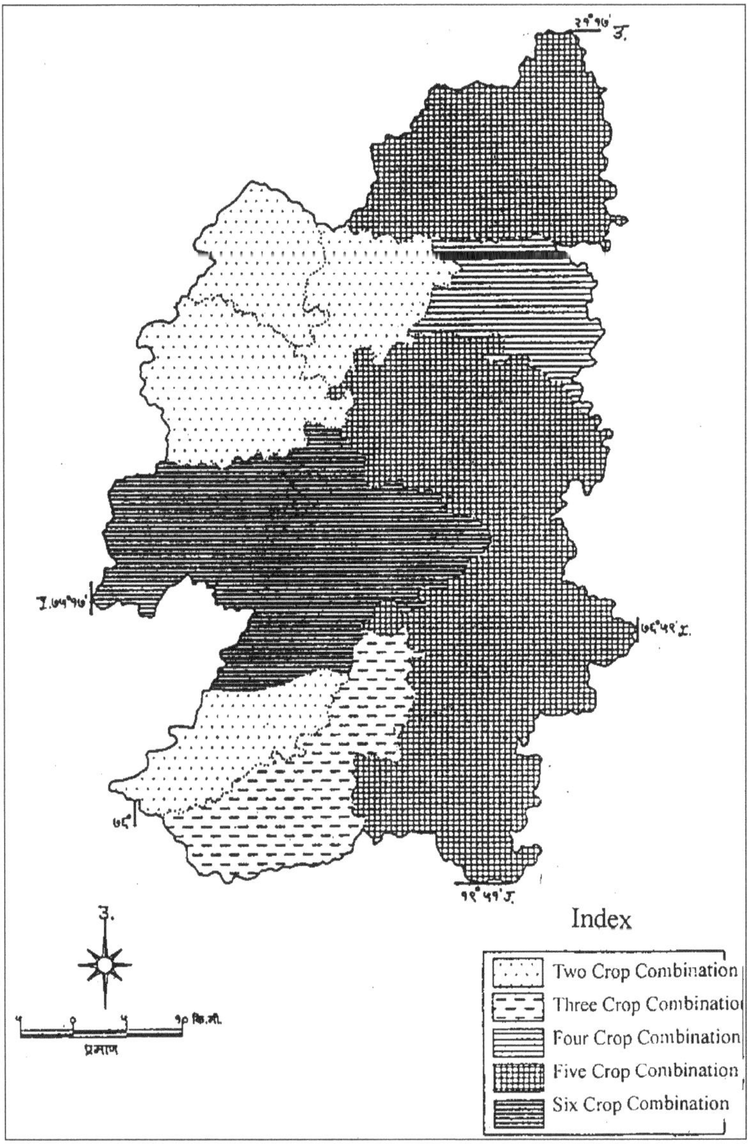

Figure 25.2: Buldhana District–Crop Combination (2000-2001)

The area under grass' grown in middle part has not consider for competing crop combination in this present study. Table 25.1 Shows ranking and crops like, Cotton, Jowar, Bajra, Tur, Soyabean, Wheat, Sugarcane, Chili, Potato in the district.

It is important to note that the environment constraints owing to geomorphic situation soils and climate have pat the limit on diversified agricultural productivity. This is reflected in the results obtained in the crop combination analysis. No. Monoculture is found in Buldhana district in the both the year. One Seventy four villages in the Nandura, Malkapur, Motola, D. Raja Tahsils have two crop combination. Among Sixth Eight Villages found cotton and Jawar in two crop combination is observed in the North western and South western part in the study area, along the puma basin. Cotton is cultivated in this area, because of black soil belt. In this study are Three crop combination is found in Sindhkhed Raja Tahsils having combination of cotton, wheat, Jawar, Pulses, and vegatable. The Shegaon Tahasils only one have dominance of four combination region, cotton, Jowar, Pulses, Tur, oilseeds, and groundwater have located in the Eastern part region in the study area. Five crop combination is identified in five tahasil sach as Jalgaon Jamod, Khamagaon, Mehakar, Lonar, Sagrampur. In this crop combination crops Cotton, Jawar, Wheat, Mug, Tur, Gram, Vegetables Oil seeds, Chill, Patato, Maize, Soybean.

It is located in the eastern part of the study area. Sloppy and plain region and black soil along the river influence the cropping pattern in this tahsils. Chikhali and Buldhana Tahasils have six crops combination regions, cotton, Jawar, Pulses, Gram, Wheat, Vegetable, Tur Bajjara, Sugar cane and maize. There are six crops amongst which for joint make a five combination in different Tahasils of Buldhana districts. The favourable climatic condition suitable rainfall, and irrigation facilities in some of western Tahasils of the district influence to cultivate varieties of crops which indicate a crop diversification.

Conclusion

A systematic analysis of crop combination regions from the point of view a dominant crop regions depicts that weaver's method is given the more as propriety results and confirm the realty in Buldhana District. Four crop combination has observed in only one

Tahasil in Shegaon in this Tahasil crop combination cotton, Jawar Pulses, Vegetable and oilseeds are cultivated on the whole, the tahasil of Buldhana District convey a sense of vigorus diversities in sectoral and spatial development.

References

Das, M.M., 1984. Agricultural land-use and cropping in Assam. *A Geographical Review of India, Kolkata* Vol. 44/4

Hussein, Majid, 1992. Crop combination region in Uttar Pradesh. *A Study of Methodology,* Vol. 44 No. 2.

Hussein, Majid, 1999. *Systematic Agricultural Geography.* Rawat Publication, New Delhi.

Kumbhar, R.A., 1978. *Agricultural Landuse in Godavari Basin.* Pune University, Pune.

Vaidya, B.C., 1990. Cropping Pattern in Yashoda Basin. *The Deccan Geographer,* Secunderabad, 28/2 and 3.

Chapter 26

Forest Management in Morena District

Vinayak Singh Tomar[1] *and N.S. Dadoriya*[2]

[1]*Head, Department of Zoology, Govt. P.G. College, Morena, M.P.*

[2]*SOS in IGAEERE, Jiwaji University, Gwalior, M.P.*

ABSTRACT

The total forest area of the Morena district is 87688.42 ha. Which is 17.38 per cent the total geographical area. This is below the norm of 33 per cent area aimed in National and state forest policies. Main hill range is vindhyan. Main geological formation are deccan and vindhyan. The division consist of 61 forest blocks out of which 18 are reserved and 43 protected forest blocks, conserving the plan metered area of 41715.01 ha and 45973.41 ha. Respectively the notified area of division in 1998 was 97120.48 ha. The difference between notified and plan metered exists since the previous working plan period.

There are several new tasks and challenges before the forests officials at present under the circumstances, it is absolutely necessary that sufficient knowledge, skill and resource are made available to them so as to enable them to execute these task and face the challenges. In view of this, new direction shall have to be given to the efforts for human resource development with a view to benefit from the latest techniques, it is necessary to adopt information technology in the forest management. A clear vision and strategy is required to fulfill the above objectives. Through this policy management of forest through a rational legal and institutional infrastructure, for conservation development and sustainable use of forest, use the ecological, economies, social and technical resources of the state will be done in such a way that apart from environmental protection ecological balance and groundwater conservation needs of forest dependent communities may also be fulfilled and productivity of forests may be enhanced so that, along with the development of forest resources, socio-economic development of these communities may be ensured by providing them employment.

It can be done by following suggestion

- ✰ Save trees as possible.
- ✰ Low amount use of fuel wood.
- ✰ Control illicit grazing.
- ✰ Use joint forest management and social forestry

Keywords: *Management, JFM, Sustainable development, Socio-economic development.*

Introduction

The total forest area of the Morena district is 87688.42 ha, which is 17.38 per cent the total geographical area. This is below the norm of 33 per cent area named in national and state forest policies. Most of the area of division is plateau. Main hill range is vindhyan. Main geological formation are deccan trap and vindhyan. The division consists of 61 forest blocks. Out of which 18 are reserved and 43 protected forest blocks, covering the planimetered area of 41715.01 ha and 45973.41 ha respectively. The notified area of division in 1990 was 97120.478 ha. The difference between notified and planimetered exists since the previous working plan period.

Madhya Pradesh has the largest forest cover in the country. The forests of the state are very important because of the abundance of biodiversity and for being the catchment area of many important rivers. Therefore, the forests of the state contribute significantly to the environmental and ecological balance and water conservation in the country. The state has a leading place in the wildlife conservation as well. About 11.4 per cent protected area (national park and sanctuaries) of the forests falls under the wildlife management. About 20 per cent of the country's and about 10 per cent of the world's population of tigers is in Madhya Pradesh. Therefore, the state has been named as the 'Tiger State'. The National Forest Policy, 1988 has given a new dimension to the forest management and protection. According to this policy, the ecological and environmental significance of forests has been kept supreme and the rights of the communities dependent on the forests have been given priority. Instead of being market-oriented, the forest management has been made sustenance-oriented for the communities dependent on forests, especially tribal women and oilier weaker sections.

Legal Status of Forest

Reserved Forests and Protected Forests of Morena

Table 26.1: Present Forest Status of Morena

Sl.No..	*Name of Area*	*Reserved Forests (ha)*	*Protected Forests (ha)*	*All Forest Area (ha)*
1.	Morena	5544.380	3926.130	9470.510
2.	Jaura	14383.923	1571.980	15955.903
3.	Pahadgarh	12099.516	39698.590	51758.106
4.	Sabalgarh	13222.509	6673.470	19895.979
	TOTAL	45250.328	51870.170	97120.498

Materials and Methods

Discussion and fixation and management circles of Morena Forest district Considering the National Forest Policy 1988, directives of the Government and local condition, the major objectives of this working planning to improve existing forest density; to increase productivity of understood and blank area after rehabilitating them and to meet the demand of the forest production of local people.

Working plan of improvement treatment management circle. In this working circle forest of 0.05 to 0.06 density are included in which between sonic small patches of well stocked forest area are included.

Table 26.2: Management Circles have been Constituted

Sl.No.	Management Circle	No. of Treatment Series		Forest Area of Management Circle (ha)	Forest Area of Management Circle (Per cent)
		Normal	Samity Based		
1.	Improvement treatment working circled	4	5	10360.84	11.82
2.	Rehabilitation working circle	18	90	74987.73	85.51
3.	Soil and water conservation working circle	1	5	2339.85	2.67
	Total	23	100	87688.42	100.00

Forest Management

Working plans for the forest management shall be prepared on scientific basis, which shall be revised from lime to lime. Management of the entire forest area shall be done only as per working plans. In the areas having vigor less root stocks because of ageing, high forests (seed originated) shall he extended through plantation, In view of the natural association of the forest dependent communities with the forests the main objective of the forest management shall not be to earn revenue, but to accord priority to the interests of these communities. Local species shall be accorded priority in the natural forest area. Micro plans of the areas under joint forest management shall be prepared within the ambit of the working plans. At least 10 per cent forest area shall be kept under intensive management. For this purpose degraded forests of good site quality shall be selected.

Joint Forest Management

Madhya Pradesh government, forest department has passed a revised resolution for ensuring participation of local villagers in the protection to ensure effective implementation of joint forest management, Semite based treatment series in respective working circles have been constituted. In Morena division 100 forest village committees are formed which constitute 23231.64 ha forest area.

Joint Forest Management (JFM) is a concept of developing partnerships between fringe forest user groups and the Forest Department (FD) on the basis of mutual trust and jointly defined roles and responsibilities with regard to forest protection and development. In JFM, the user (local communities) and the owner (Government) manage the resource and share the cost equally, however it is difficult to generalize the JFM concept and approach in the light of variations across the nation with respect to geography, resource base, socio-economic status, cultural diversity and pressures on forests.

JFM is slowly emerging into a form of sustainable forestry, which augments the forestry regime with processes for rapid adaptation to changes in what people need, want, and can do. As an adaptive social process it is striving to create sufficient future forest production opportunity to satisfy potentially competitive/conflicting interests that would diminish the forest is left unresolved. The challenge with JFM has thus been to develop social processes that recognize, accommodate, and respond more effectively to diverse and dynamic perspectives of what the forest is about and should be.

Forest Extension and Social Forestry

To fulfill forest produce based needs, plantation on community/ private lands, on bounds on agriculture fields, are necessary, non conventional sources of energy should also used for this local villagers should be imparted with the knowledge of forestry extension, well trend and dedicated staff should deputed for social forestry programmes.

Management of Non-Timber Forest Products

The importance of forest of non-timber forest produce is increasing. Tendu produce is being collected through co-operative societies and a major source revenue, it is necessary to given training to villagers increase production tendu patta. Apart from tendu patta, grass, gum and medicinal plant are collected. Forest staff and villagers are required to be given prop training identification of medicinal plants and their collection techniques, *ex situ* and *in situ* conservation of medicinal plant should be done.

Miscellaneous Regulation

Maintenance of boundaries and boundary pillar; condition of boundary and boundary pillars is very bad, pillars are seen occasionally in the forest areas, boundaries of many forest blocks are not clear.

Plantation

Most of the old plantation are failure, new plantations must be done according to site specific project report. High-tech plantation will be done as per the instructions of P.C.C.F. Madhya Pradesh.

Conclusions

The forest of the Sabalgarh block of Morena division is dry deciduous type-under line row does not always growth of canopy. The whole of the area is covered with Cardhai (*Anogesis pendula*) and Kher (*Acacia catechu*). Inter spers species *Gigipher morosiena, Dieya speros* and *Butia monosperm*. Low rain fall, high temperature and desiccating winds affect the area adversary. There is continuous the short of moisture in the area. That does not always growth and rocky substratum does not always growth. Under the grazing pressure and illicit falling at the poor regeneration in the area. This area need to be close minimum for a period of 3 years. Grazing should be totally stopped and where ever soil in unliving. Seeds of kher should be sown, this will always conservation of moisture and regeneration. Throw seeds and woods protected on area returning. It denuded refectrisite. Joint forest management communities are just in the name sap. Since this committees are most getting any production. There is not interested protection in the also. It would be birth wile in the JFM committees in function on the grout forest area.

References

Allen, T.F.H. and Hoekstra, T.W., 1994. Toward a definition of sustainability. In: *Sustainable Ecological Systems: Implementing an Ecological Approach to Land Management*, (Eds.) W.W. Covington and L. DeBano. Rocky Mountain Forest and Range Experiment Station, US Department of Agriculture.

ITTO, 1998. Criteria and indicators for sustainable forest management of natural tropical forests. ITTO Policy Development Series No. 7, International Tropical Timber Organization, Japan.

Saxena, N.C., 1997. *The Saga of Participatory Forestry Management.* CIPOR, Jakarta, Indonesia.

Tucker, R.P., 1993. Forests of the Western Himalaya and the British Colonial System (1815–1914). In: *Indian Forestry: A Perspective*, (Ed.) A.S. Rawat. Indus Publishing.

United Nations, 1992. Non-legally binding authoritative statement of principles for a global consensus on the management, conservation and sustainable development of all types of forests. Report of the UN Conference on Environment and Development, Rio de Janeiro, 3-14 June, Vol. 3.

World Commission on Environment and Development. 1987. *Our Common Future: The Bruntland Report*. Oxford University Press from the World Commission on Environment and Development, New York.

Index

A

Acid affected 162
Acid deposition 266
Acid rain 262, 265
Afforestation 116, 217
Agricultural development 164
Agricultural practices 72
Agricultural production 182
Agricultural products 190
Agricultural system 27
Agriculture 27, 62, 112
Agriculture and food security 62
Agriculture climate change 29
Agriculture environment 140
Agriculture-irrigation 46
Agro-climatic zones 70
Animal production 64
Anthropogenic costs 84
Antropogenic activities 254
Applied fertilizers in 156
Aquatic animals 246
Artificial recharge 48
Asola-Bhatti wildlife sanctuary 120
Atmospheric concentrations 72

B

Bhakra Storage 132
Bio-diesel 33
Biodiversity 111, 246
Biological communities 247
Biological influents 203
Breathing space 105
Business environment 10, 11, 12, 15

C

Canarium 2
Canarium bengalense 1
Canarium resiniferum 1

Carbon capture and storage (CCS) technology 103
Carbon taxes 93
Cataracts 265
Central Ground Water Board (CGWB) 231
Central Public Health and Environmental Engineering 132
Centre for Science & Environment 11
CFC-13 262
CFC's 263
CH_4 262
Chambal river 164, 258
Chlorofluorocarbons (CFC) 262, 265, 263
Cholera 55
Climate change 10, 27, 60, 61, 63, 64, 65, 70, 72, 73, 74,77, 84, 86, 93, 95, 108, 110, 114, 140, 177, 178, 180, 182, 219, 229, 262, 264
Climate factors 71
Climatic conditions 41
Climatic variability 72
CMSA 76
CO_2 29, 30, 71, 101, 103, 109, 115, 141, 142, 184, 221, 224, 262, 263
Complex 12
Conservation 116, 124, 282
Controlling erosion 282
Cotton 289, 291
Crocodylus palustris 248
Crop combination analysis 285
Cropping pattern 48, 268, 271, 285
CSO 231
Ctro coagulati 58

D

Database 288
Deforestation 212
Degraded forest 162
Delhi Metro Rail Corporation (DMRC) 127
Diarrhoea 55
Dietary diversification 68
Disaster 262
Dissolved organic carbon (DOC) 221
District ground water broachers 231
Double cropping 275
Drainage 41
Drinking water 54
Dysentery 55

E

Earth's vegetative cover 1
Economic environment 19
Economic goal 154
Economic impacts 35
Economic policy 24
Economical environment 15
Economy 114
Efficient management analyses 10
Egg-plants 173
Electro-coagulations 58

Electro-dialysis 58
Energy strategy 184
Environment 143, 192, 203
Environment scanning 10
Environmental effects 72
Environmental stability 116
Eroded lands 282
Erosion 217, 280
Ex situ 58
Exposition 288
External forces 11

F

FAO 61
Farmers' institutions 75
Farming system 165
Financial support systems 76
Fishing 258
Floods 63, 113
Food and nutrition security 60, 63, 65
Food and water security 65
Food security 61, 73
Forest 121
Forest based small-scale industry 174
Forest extension 297
Forest management 293, 296
Forest resource 1, 9, 116, 118
Freshwater 149
Freshwater ecosystems 247

G

Gangetic dolphin 251
Gavialis gangeticus 248
GDP 87, 90, 101, 104, 179, 184
General forces 11
Genetic erosion 77
GHG 29, 94, 95, 103, 104, 221, 227, 236
Global warming 109, 262
Globalisation 23, 24
GPS device 249
Green accounting 151
Green leafy vegetables 68
Green lung 116
Greenhouse effect 262
Greenhouse gas emissions 177
Greenhouse gases products 115
Groundwater 48, 54
Groundwater contamination 54, 56
Groundwater depletion 42
Groundwater level 40
Groundwater resource 39

H

Habitat features 249
High yielding varieties (HYVs) 238
Himalayan watershed 187
Household food security 61
Human development report (HDR) 220
Human economic activities 191
HYV animals 175

I

ICRISAT 76
Immobilization technologies 59

In situ 58

Indian agriculture 70

Indian Council of Agricultural Research 32

Individual food security 61

Integrated pest management 33

Intergovernmental Panel on Climate Change (IPCC) 29, 71, 84, 94, 101

International year of freshwater 55

IUCN 249

J

JFM 294

Jhum (Shifting) cultivation 2, 6

Joint Forest Management 296

K

Kansai Paint Company 13

Karbi 2

Kharif 271

Kishau dam 133

Kishu dam 137

L

Lakhwar dam 133

Land degradation 77, 209

Land-Use 268

Land-Use Pattern 159, 269

Legal environment 18

Liberalisation 23, 24

Livelihood 73

Livelihood security 76

Livelihood sustainability 164

Livestock 165

M

Medicinal plants 175

Medicinal properties 1

Methane 29, 262

Methane emissions 180

Microbial biosorption 58

Migratory insects 144

Milk production 174

Mitigation policies 240

Moisture index 234

N

Narsingh Danda area 204

National aquatic animal 251

National Centre for Organic Farming (NCOF) 76

National Chambal sanctuary 246

Natural capital stock 151

Natural environmental resources 147

Natural resin 1

Natural resources 150, 187, 195, 215, 228

Natural resources degradation 207

Nerolac Company 13

New Delhi Municipal Council (NDMC) 132

New Economic Policy 1991 22

NGO 125

Nitrous oxide 29, 262

Non-halogenated volatile organic compounds (VOCS) 58

Non-timber forest products 297
Northern ridge 119
NOx 262
NREGA 76
NSSO 231
Nutrient 80
Nutrient management 79
Nutrition 90
Nutrition security 61

O

OECD 94, 95, 101
Off-season vegetables 172
Oilseeds 32, 286
Ozone depletion 262

P

Parallel lined canal 135
Pathogens 140
Patiala and East Punjab States Union (PEPSU) 283
Pedalfers 282
Pedocals 282
Pests 140
Photochemical oxidation 58
Physiography 41, 43
Phytoremediation 58
Plant disease management practices 142
Plant root growth 80
Plantation 298
Platanista gangetica gangetica 248
Political environment 17
Poverty 149
Privatisation 23
Protected forest 120, 295
Providing Regional Climates for Impact Studies (PRECIS) 227
Pulses 32, 286, 289, 291

R

R&D 96, 101
Rabi 79, 273
Rainfall 37, 41, 71
Rajasthan desert 283
Raw water 135
Reform measures 21
Regional food security 61
Relativity 12
Remediation 54, 57
Renuka dam 133
Reserve forest 118, 295
Resin 2
Resin tree plantation 9
Rice 32, 170
Ridge forest 118
Rising sea levels 63

S

Safe drinking water 56
Sal 2
Salt affected 162
Sea level rise 183
Second assessment report (SAR 220
Social environment 16
Social forestry 297
Socio-economic development 294

Soil 79
Soil conservation 280
Soil conservation associations 284
Soil degradation 203
Soil erosion 278
Soil infertility 171
Soil moisture 79, 80
Soil resources 156
Soils 277
Southern ridge 120
Stabilization solidification 59
State groundwater profiles 231
Stratospheric ozone depletion 264
Subsistence cereal crops 165
Surface water resources 43
Sustainable agriculture 74, 80, 156, 163
Sustainable agriculture development 285
Sustainable development 9, 35, 47, 146, 147, 151, 152, 187, 294
Sustainable livelihood 165, 172

T

Technological environment 19
Tehri Dam 132, 137
The United States Environmental Protection Agency (USPA) 58
Tiger state 295
Topography 43
Traditional agricultural farming 167
Tree cover 121, 123
Tropospheric ozone 72

U

UN Climate Change Conference 103
UN Development Programme (UNDP) 220
Under-Nutrition 65
UNFCCC 177, 179
UNO 216
Upper Sindh Basin 164

V

Vegetable crops 286
Vegetables 175
Vegetation types 233
Vulnerability Index 234

W

Water balance 233
Water depletion 77
Water erosion 162
Water resources 35, 45, 114
Water temperature 249
Water use efficiency 79, 80
Water-borne infections 55
Weather disasters 113
Wildlife 111
Wind erosion 162
World agriculture 27
World conservation strategy 147

Y

Yamuna 117

www.ingramcontent.com/pod-product-compliance
Ingram Content Group UK Ltd.
Pitfield, Milton Keynes, MK11 3LW, UK
UKHW022214051026
15624UKWH00001B/150

9 788170 35994